Digital Logic Techniques

principles and practice

TUTORIAL GUIDES IN ELECTRONIC ENGINEERING

Series editors
Professor G.G. Bloodworth, *University of York*
Professor A.P. Dorey, *University of Lancaster*
Professor J.K. Fidler, *Open University*

This series is aimed at first- and second-year undergraduate courses. Each text is complete in itself, although linked with others in the series. Where possible, the trend towards a 'systems' approach is acknowledged, but classical fundamental areas of study have not been excluded, neither has mathematics, although titles wholly devoted to mathematical topics have been eschewed in favour of including necessary mathematical concepts under appropriate applied headings. Worked examples feature prominently and indicate, where appropriate, a number of approaches to the same problem.

A format providing marginal notes has been adopted to allow the authors to include ideas and material to support the main text. These notes include references to standard mainstream texts and commentary on the applicability of solution methods, aimed particularly at covering points normally found difficult. Graded problems are provided at the end of each chapter, with answers at the end of the book.

1. Transistor Circuit Techniques: discrete and integrated — G.J. Ritchie
2. Feedback Circuits and Op. Amps — D.H. Horrocks
3. Pascal for Electronic Engineers — J. Attikiouzel
4. Computers and Microprocessors: components and systems — A.C. Downton
5. Telecommunication Principles — J.J. O'Reilly
6. Digital Logic Techniques: principles and practice — T.J. Stonham

Digital Logic Techniques

principles and practice

T.J. Stonham
Department of Electrical Engineering
Brunel University

VNR (UK) **Van Nostrand Reinhold (UK) Co. Ltd**

First published in 1984 by
Van Nostrand Reinhold (UK) Co. Ltd
Molly Millars Lane, Wokingham, Berkshire, England

Reprinted 1985

Typeset in Times 10 on 12pt by Colset Private Ltd, Singapore

Printed and bound in Hong Kong

Library of Congress Cataloging in Publication Data

Stonham, T.J.
Digital logic techniques.

(Tutorial guides in electronic engineering)
Includes index.
1. Logic design. 2. Digital electronics. I. Title.
II. Series.
TK7868.L6S76 1984 621.3815'3 84-5171
ISBN 0 442-30595-8 (pbk.)

Preface

The developments that have taken place in digital electronics over the past twenty years have no parallel in any other branch of engineering. A striking fact is that the real cost of hardware has halved every year whilst its complexity has, on average, quadrupled every three years. This is clearly illustrated by considering semiconductor memory. In 1964 one bit of storage would have comprised a single JK flip-flop constructed out of discrete components and cost the equivalent of twenty hours of a graduate engineer's time. Today single chips contain over a quarter of a million bits of storage and have a cost equivalence of a tenth of a second of an engineer's time per bit. The increase in complexity and decrease in cost are both of the order of one million since the mid 1960s. Equivalent progress in, for example, the motor industry would have provided us with luxury cars requiring only half a gallon of petrol for life having sufficient thrust to go into orbit and a price tag of about twenty pence.

With these developments as a backcloth, this introductory book aims to provide the necessary fundamentals of digital logic, whilst familiarising the student with design aspects and techniques at the system level. This is most relevant as current technology dictates that digital systems will be constructed from complex components, widely referred to as registers.

The starting point of the text is an objective comparison between analogue and digital representation of data. The student engineer must be aware that any given information processing task can be implemented either with analogue or digital electronics and in optimal design neither approach can be totally ignored. Digital methods of data representation are established in the opening sections. In line with the author's intentions to introduce the reader to contemporary aspects of the subject, the concepts of error detection and correction are included.

Boolean algebra provides a mathematical framework for digital electronics and has been used to link the propositional description of an operation to its implementation as a gate-level logic circuit. Combinational logic design from first principles provides the student with a vehicle to develop his understanding of digital electronics. Traditional methods of logic minimisation have been included as they are convenient 'pencil and paper' techniques. The emphasis placed upon them is, however, not high — reflecting their lessening importance in modern logic design. Cellular logic is presented as an alternative structure to canonical forms and is more relevant to VLSI implementation.

Sequential logic design is founded upon the flip-flop as an elementary building block. A strategy for the design of finite state machines based upon a generalised sequential logic structure comprising three data sets — the inputs, the internal states and the outputs — and two logic systems — the next state logic and the output logic — has been developed. Students can thereby create special-purpose sequential systems and also acquire a sound foundation for more advanced studies in Automata Theory.

The latter part of the book concerns the digital system as an entity. The implementation of logic systems in programmable hardware such as memories and arrays is examined and the specification of a system as a set of registers interconnecting via a bus structure is outlined. Fault diagnosis in a digital system has been defined and elucidated. This has often been regarded as a research topic but is now assuming increased importance in digital design. The author feels that these objectives must be introduced to the student engineer.

At the end of any engineering activity an artifact is produced. It is therefore essential that due attention is paid to the electrical properties of logic components. A comparison of the principal logic families is presented and the student's attention is drawn to potential problems which can arise when hardware systems are constructed.

When studying any subject, the student must experience for himself the fundamental concepts. To this end, numerous worked examples and marginal notes have been included in the text and graded problems with solutions given at the close of each chapter.

The author wishes to thank his Consultant Editor, Professor J.K. Fidler, for his valuable advice throughout the preparation of this text. Thanks are also due to Professor I. Aleksander and colleagues at Brunel University for counsel and encouragement and to Mrs. E. Flanagan for invaluable secretarial assistance. Finally, thank you to the Electrical Engineering students at Brunel for being a 'test-bed' for the approach to digital systems adopted in this book.

Contents

Numerical Representation of Information

1

Objectives

- ☐ To distinguish between analogue and digital quantities.
- ☐ To investigate the binary number system.
- ☐ To define and examine binary coded decimal numbers.
- ☐ To specify the essential properties of a position sensing code.
- ☐ To introduce the concepts of error detection and correction in binary data.

In almost all activities, we are constantly dealing with quantities or measurements. This information is expressed in the form of numbers and can be processed in a digital system, provided it can be represented in an electronic form.

Information is almost always encoded in binary in a digital system and a wide range of codes exist. The choice of a particular code is influenced by the type of operation to be carried out on the data.

In this chapter, methods of representing numerical data in binary will be introduced. It is essential that the designer is familiar with fundamental coding techniques, as the form in which information is represented has significant influence on the design, performance and reliability of a digital system.

Analogue and Digital Data

The first step in any data processing operation is to obtain information about the objects or phenomena of interest. The acquisition of information usually involves taking measurements on some property or characteristic of a system under investigation. In order to evaluate and assess the system, the measurements are monitored, scaled, compared, combined, or operated on in various ways. It is therefore essential that we have standard ways of representing our information.

Measurements can be divided into two broad categories. An analogue measurement is continuous and is a function of the parameter being measured. Conversely, a digital quantity is discrete and it can only change by fixed units.

Figure 1.1 shows two beakers being filled, one from a dripping tap and the other from a trickling tap. The build-up of water in the first beaker is typically digital, as the smallest change in volume is equal to one drip and the value increases in steps. In the second beaker, the volume of water increases continuously with time. It has an analogue property.

All drips are assumed to be the same size.

Analogue and digital electronic circuits can be devised to process data. Analogue circuits such as operational amplifiers are cheap and powerful, but the analogue approach has two main disadvantages. The real world problem must have a precise electronic model and accurate measurements need to be made on that model.

For details on Op-amps, refer to Horrocks, D.H. *Feedback circuits and Op-amps* (Van Nostrand Reinhold, 1983).

In digital electronic circuits, there is no need to make precise time-dependent measurements as a problem is modelled by a set of rules based on logic, which will

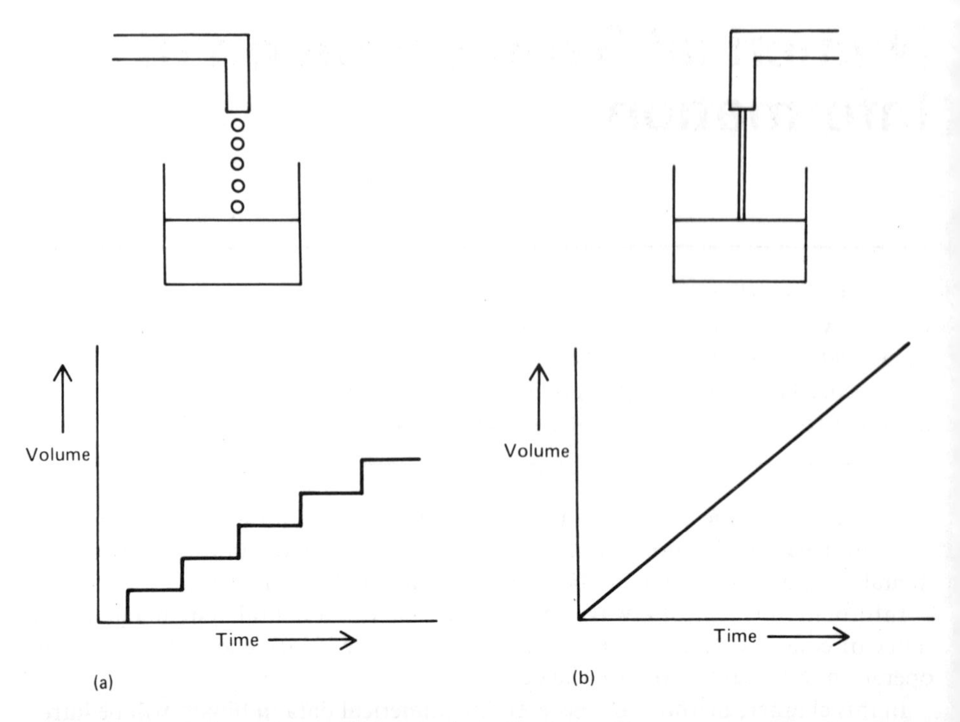

Fig. 1.1 (a) A digital or discrete system. (b) An analogue or continuous system

The resolution of a digital system determines the smallest detectable change in its parameters. It is dependent on the number of bits used to represent quantities.

be developed throughout this book. One major drawback of digital circuits is resolution. If the resolution is increased by reducing the minimum step changes allowed in the parameters, the amount of electronic circuitry must increase.

The electronics engineer should be aware that analogue, digital or a hybrid combination of both methods may be relevant to a particular information processing operation. Neither approach should be precluded as any one method may provide a significantly simpler solution to a given problem.

Information is all about numbers. A number is a label with a unique meaning, and the remainder of this chapter will examine ways of representing information.

Exercise 1.1 Which of the following items involve analogue or digital quantities?

(1) Traffic flow
(2) Battery voltage
(3) Telephone exchange
(4) Temperature
(5) Vehicle speed
(6) Size of a pumpkin
(7) Pips in a pumpkin
(8) A number

Number Systems

The Decimal Number System

A number system uses a set of symbols, known as digits. The total number of different symbols used by a given number system is its base or radix. In the decimal umber system, ten symbols are used. They are 0, 1, 2, 3, 4, 5, 6, 7, 8, and 9 giving he base of the system as 10. The symbols are ordered and a change between two djacent numbers is a unit or quantum. A complete number is the weighted sum of he powers of the base. Take, for example, the number 471. The base raised to ower 0 is weighted by 1, (the right-hand symbol). The base to power 1 is weighted y 7 and so on. Hence the number is equal to

$$(4 \times 10^2) + (7 \times 10^1) + (1 \times 10^0)$$

he symbols themselves have a value and there is also a value associated with their ositions. The complete number is the sum of the product of each digit value and its ositional value. A general expression for the magnitude of a decimal number is

$$D = d_n 10^n + d_{n-1} 10^{n-1} + \ldots + d_1 10^1 + d_0 10^0 + d_{-1} 10^{-1} + \ldots + d_{-n} 10^{-n} \quad (1.1)$$

nd the number is represented by the symbol string

$$d_n \, d_{n-1} \ldots\ldots d_1 \, d_0 . d_{-1} d_{-2} \ldots d_{-n}$$

here $d_i \in (0,1,2,3,4,5,6,7,8,9)$

$\in$ reads 'belongs to'.

he decimal number system no doubt evolved because humans have 10 fingers and the early mechanical computers and calculating machines attempts were made at sing base 10 numbers. Unfortunately there are very few devices that have ten atural states, one for each decimal digit. In contrast, there are many systems here two discrete states can be detected without requiring precise measurement. ome typical binary state pairs are on/off, present/absent, high/low, true/false, ositive/negative, stop/start, to quote a few.

'digit' is the latin word for 'finger'.

Because of the ease with which two distinct states can be represented, the base 2 binary number system has been universally adopted for electronic digital pro-ssing systems.

he Binary Number System

the binary number system, there are only two symbols, 0 and 1, and the base is erefore 2. The value of a pure binary number is given by the polynomial

$$B = b_n 2^n + b_{n-1} 2^{n-1} + \ldots + b_1 2^1 + b_0 2^0 + b_{-1} 2^{-1} + \ldots + b_{-n} 2^{-n} \quad (1.2)$$

Numbers of the form shown in Equation 1.2 are called pure or natural binary numbers. Other binary codes exist that do not have this polynomial form.

d the number is written as the string

$$b_n \, b_{n-1} \ldots b_1 \, b_0 . b_{-1} b_{-2} \ldots b_{-n}$$

here $b_i \in (0,1)$ and is a binary digit or BIT.

$b_i \in (0,1)$ means b_i is either 0 or 1.

onversion Between Decimal and Binary Numbers

given quantity can be represented by a number in any base. The digit strings will ffer, depending on the base being used, but the magnitude of the numbers will ways be identical. Hence, given a number in base a, there is one and only one mber in base b that has the same magnitude.

The most important bases are two, which is used in digital systems, and ten, ich is used by man in everyday communications. A binary number can be con-rted into a decimal number by evaluating its polynomial, given in Equation 1.2.

Worked Example 1.1 Convert 11011_2 into decimal.

The polynomial of the binary string is

See Equation 1.2.

$$(1 \times 2^4) + (1 \times 2^3) + (0 \times 2^2) + (1 \times 2^1) + (1 \times 2^0)$$

The decimal value of the polynomial is

$$16 + 8 + 0 + 2 + 1 = 27_{10}$$

Hence $11011_2 = 27_{10}$

In order to convert from decimal to binary, a successive division procedure is used. In the general case, decimal number D is identical to binary string

$$b_n b_{n-1} \ldots b_1 b_0$$

Hence

$$D = b_n 2^n + b_{n-1} 2^{n-1} + \ldots + b_1 2^1 + b_0 2^0 \tag{1.3}$$

Dividing each side of Equation 1.3 by the new base (which is 2) gives

$$\frac{D}{2} = b_n 2^{n-1} + b_{n-1} 2^{n-2} + \ldots b_1 2^0 + b_0 2^{-1} \tag{1.4}$$

The left-hand side of Equation 1.4 consists of an integer part and a remainder where

$$\text{Integer} \left| \frac{D}{2} \right| = b_n 2^{n-1} + b_{n-1} 2^{n-2} + \ldots + b_1 2^0$$

and $$\text{Remainder} \left| \frac{D}{2} \right| = b_0$$

Note the order of the bits. The 1st division gives the least significant bit, i.e., the rightmost bit.

The least significant bit of the binary string is therefore the remainder after the first division.

A further division of Integer $\left| \frac{D}{2} \right|$ by 2 will give b_1 and the process can be repeated until the complete binary string has been calculated.

In general, when converting from a smaller to a larger base use a polynomial expansion (see Worked Example 1.1). When converting from a larger to a smaller base use successive division (Worked Example 1.2)

Worked Example 1.2 Convert 19_{10} into binary.

$$D = 19$$

2 is the base of the new number.

Divide by 2 $\quad \frac{D}{2} = \frac{19}{2} = 9 \text{ Rem } 1 \qquad \therefore b_0 = 1$

Rem means remainder.

2nd Division $\quad \frac{\text{Int} \left| \frac{D}{2} \right|}{2} = \frac{9}{2} = 4 \text{ Rem } 1 \qquad \therefore b_1 = 1$

3rd Division $\quad \frac{4}{2} = 2 \text{ Rem } 0 \qquad \therefore b_2 = 0$

4th Division $\quad \frac{2}{2} = 1 \text{ Rem } 0 \qquad \therefore b_3 = 0$

5th Division $\frac{1}{2} = 0$ Rem 1 $\therefore b_4 = 1$

Hence $19_{10} = 10011_2$

Note the order of the bit string.

A binary number Equation less than 1 can be converted to decimal by calculating its polynomial (Equation 1.2). A decimal number less than 1 can be converted to binary by a successive multiplication of its fractional part by the new base (2). The integer part of the resulting decimal number after each multiplication, is a coefficient of the binary number, starting with the most significant bit. If a decimal number having both an integer and a fractional part, is to be converted to binary, both parts must be processed separately. Use successive division for the integer part and successive multiplication for the fractional part.

Non-Pure Binary Codes

Binary Coded Decimal

A binary number system always uses only two different symbols, but the value of the numbers need not be based on a polynomial series. In an electronic system, the information is in binary form but when it is output to a user, decimal is more convenient.

A form of coding that enables a very simple conversion between binary and decimal is binary coded decimal or BCD. In this case, each digit of a decimal number is replaced by 4 bits corresponding to its pure binary value.

Worked Example 1.3

Convert 45_{10} to BCD

D = 45

Replace each digit with 4 bits binary

$D = 0100/0101$

Hence $45_{10} = 1000101_{BCD}$

Leading zeros from most significant digit may be ignored.

To convert from BCD to decimal, the binary word is divided into groups of 4-bits starting from the least significant bit and the decimal equivalent of each group obtained.

Worked Example 1.4

Convert 1010011_{BCD} to decimal

Divide BCD into groups of 4 0101/0011

Decimal value of each group 5/3

Hence $1010011_{BCD} = 53_{10}$

Start from LSB. If the most significant group has less than 4 bits assume leading zeros.

BCD coding results in input and output devices being very simple as the value of each decimal digit is dependent on only 4 bits. The arithmetic process however becomes more complex as the rules of binary arithmetic no longer apply and certain pure binary numbers are not allowed. The above BCD coding is sometimes

If a code is simply referred to as BCD, assume that it is 8421 weighted, i.e., it uses pure binary values of the decimal digits.

referred to as 8421 BCD, as this is the positional value of the bits within each group of 4.

Many other forms of BCD exist. A decimal digit need not necessarily be coded in pure binary, but it will always require 4 bits. A total of 16 different values (2^4) can be represented on 4 bits and 10 values are required to encode the set of decimal digits. In theory there are $\frac{16!}{(16-10)!}$ different ways of selecting 10 codewords from 16. This amounts to approximately 3×10^{10} different BCD codes. The vast majority of these codes are quite random, and less than a hundred are weighted codes where each bit has a fixed weight. Some of the more common weighted codes are (4,2,2,1), (7,4,2,1) and (5,2,1,1). The weights need not be restricted to positive values and codes having some negative weights exist.

The weighting indicates the value of the corresponding bit in the group of 4 bits representing the decimal digit.

Some examples are (7,4, −2, −1) (8,4, −2, −1) (−3,6,4,2). Code values for the decimal digits are shown in Table 1.1

The weights cannot *all* be negative.

Table 1.1 Some Weighted BCD Codes

Decimal	8421	4221	7421	5211	74-2-1	84-2-1	-3642
0	0000	0000	0000	0000	0000	0000	0000
1	0001	0001	0001	0001	0111	0111	1010
2	0010	0010	0010	0011	0110	0110	0001
3	0011	0011	0011	0101	0101	0101	1100
4	0100	0110	0100	0111	0100	0100	0010
5	0101	0111	0101	1000	1010	1011	1101
6	0110	1010	0110	1001	1001	1010	0100
7	0111	1011	0111	1011	1000	1001	1110
8	1000	1110	1001	1101	1111	1000	0101
9	1001	1111	1010	1111	1110	1111	1111

Gray Code

Quite often a digital circuit is used in a control system where the position of an object has to be measured. This can be done by encoding positions on the object and detecting the nearest position value. The edge of a conveyor belt, for example, could be encoded in binary using black and white markings as shown in Fig. 1.2. As

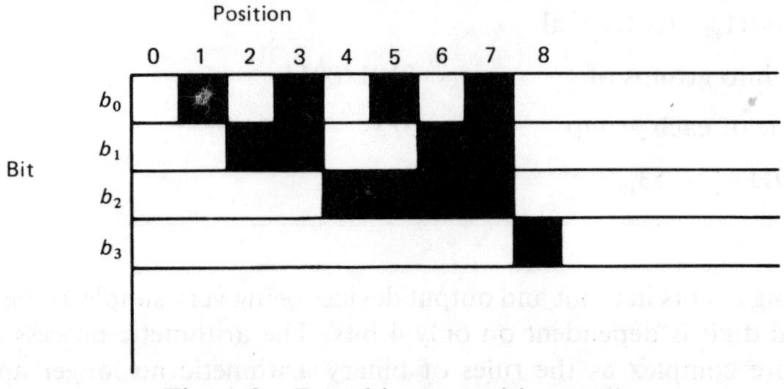

Fig. 1.2 Pure binary position coding.

the belt moves under a fixed transducer its position label can be read. If the positions are encoded in pure binary, errors can occur when the binary values representing adjacent positions differ by more than one bit.

Consider the belt moving from positions 1 to 2. Bit b_0 must change from 1 to 0 simultaneously with b_1 switching from 0 to 1. Electronic circuits can respond to changes of data within a microsecond and in practice b_0 and b_1 will not both change within this time. Either bit b_0 will change slightly before or after b_1 and erroneous position values 0 or 3 will be read when passing from positions 1 to 2. Similar errors will occur wherever adjacent positions differ by more than one bit, such as 3 to 4, 5 to 6, 7 to 8, and so on.

Switching is dependent on the alignment of the reading transducer, its linearity and the accuracy of the position code printing.

The problem can be overcome by using a reflected binary code, where adjacent values differ by only one bit. The most popular of these codes is the Gray code, which can be built up by progressively setting and reflecting the bits of the code word.

Starting from 0_{10} which is 0000_G, the least significant bit is set to 1 giving

$$1_{10} = 0001_G$$

For 2_{10} and 3_{10}, bit 1 is set to 1 and bit 0 reflected, giving

$$2_{10} = 0011_G$$
and $$3_{10} = 0010_G$$

For $4_{10} - 7_{10}$ bit 2 is set and bits 1 and 0 reflected and so on. The Gray code for decimal 0 to 16 is given in Table 1.2. It can be seen that adjacent values differ by only 1 bit. This makes the Gray code ideal for position sensing as the problems associated with the need for simultaneous changes of two or more bits never arises. A Gray position sensing code is shown in Fig. 1.3.

Table 1.2 The Gray Code

Decimal	Gray Code	
0	0	
1	1	
2	11	Set bit 1. Reflect bit 0
3	10	Set bit 2. Reflect bits 1 and 0
4	110	
5	111	
6	101	
7	100	
8	1100	Set bit 3. Reflect bits 2, 1 and 0
9	1101	
10	1111	
11	1110	
12	1010	
13	1011	
14	1001	
15	1000	
16	11000	Set bit 4. Reflect bits 321 and 0

Imagine a mirror is placed under the reflected bits.

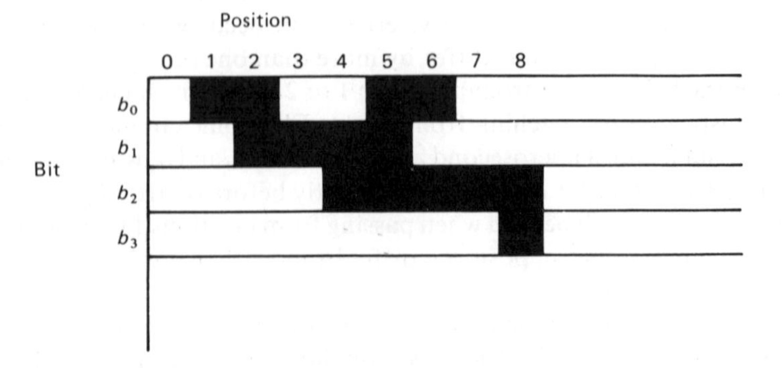

Fig. 1.3 Gray position coding.

Error-Detecting Codes

When large amounts of data are being generated and processed in a digital system, there is always the possibility of some bits being corrupted. A 1 can be changed to a 0 if part of an electronic circuit is not working, and errors can be introduced if data has to be transmitted over large distances by cable or radio.

Parity Codes

A simple method of error detection is to use an additional bit, called the parity bit, with each binary word. Odd and even parity coding can be used. In odd parity coding, the parity bit is 0 if the total number of bits set to 1 within the word, is odd. If even, then the parity bit is set to 1 so as to make the total number including the parity bit, odd.

The parity bit is generated before any data word is transmitted and a parity check can then be carried out in any subsequent part of the system. This involves a new calculation of the parity of the data word (excluding the parity bit). A comparison of the result with the parity received with the data word is then made and if they agree there is no single error in the data word.

Consider the binary word 010001. If it is to be transmitted with odd parity, the parity bit must be set to 1, giving

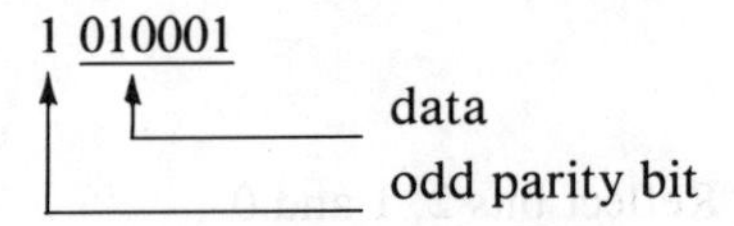

Now suppose bit b_2 is corrupted and becomes 1. The received word will be

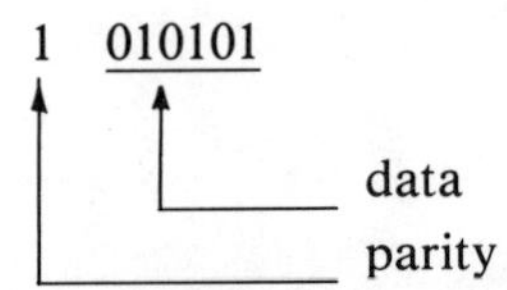

A parity check on the data would now give a value 0, but the parity received is 1. This discrepancy can be detected and indicates that an error is present in the data.

Unfortunately, with this simple parity scheme there is no way of telling where the error is. Furthermore, if two errors occurred within the same word, they would not be detected.

Even parity can also be used. The principle is the same as the odd parity scheme, except that the parity bit now makes the total number of bits set to 1 in the word even. Odd and even parity codes of pure binary numbers are given in Table 1.3.

Table 1.3 Even and Odd Parity Coded Binary Numbers

Binary	Even parity binary	Odd parity binary
0000	00000	10000
0001	10001	00001
0010	10010	00010
0011	00011	10011
0100	10100	00100
0101	00101	10101
0110	00110	10110
0111	10111	00111
1000	11000	01000
1001	01001	11001

'm out of n' Codes

An alternative method of detecting errors is to use an '*m* out of *n*' code where *n* represents the total number of bits in a binary word, of which *m* must be set to 1. If more or less than *n* bits are set to 1 then errors are present. The error-detection circuitry has to count the number of bits set to 1 in a word and compare it with *m*. This is a relatively simple operation.

Worked Example 1.5

Devise a '2 in 5' code to represent the decimal digits 0 to 9. Each codeword must have 2 bits set and be 5 bits long. Valid codewords can be identified by counting in pure binary and using only those words that have 2 bits set.

Count	Action
00000	ignore
00001	ignore
00010	ignore
00011	valid code equivalent to 0_{10}
00100	ignore
00101	valid code equivalent to 1_{10}
etc.	

The full code is

'2 in 5'	Decimal
00011	0
00101	1
00110	2
01001	3
01010	4
01100	5
10001	6
10010	7
10100	8
11000	9

Exercise 1.3 Devise a '3 in 5' code for decimal digits.

Error-Correcting Codes

Error-detection signals can be used to stop any operation in a digital system and thereby prevent any false results or operations being generated. If however the errors can be detected and corrected, the processing operations can continue.

The principles of error detection can be extended to enable errors to be both detected and automatically corrected. Two elementary techniques will now be examined, which enable single errors to be corrected.

To detect and correct a single error within a data word, the number of parity bits must be increased. Consider a 4-bit binary word

$$b_3b_2b_1b_0$$

Parity bits need to be generated for each group of 3-bits giving

P_1 parity for $b_3b_2b_1$
P_2 parity for $b_3b_1b_0$
P_3 parity for $b_2b_1b_0$

The total information which has to be transmitted is therefore

$$P_3P_2P_1b_3b_2b_1b_0$$

and in any subsequent error correction, all parity bits are recalculated and compared with the received values. Each bit in the received word can be corrupted and its binary value changed, giving a possibility of seven different single errors, which can occur in either the data or the parity sections. Supposing bit b_0 is corrupted. The parity check values of P_2 and P_3 will not then agree with the received values. A failure on parity bits P_2 and P_3 will only happen if an error occurs on b_0.

The data is binary and therefore there are only two possible values for each bit,

Table 1.4 Error Locations on a 4-bit Data Word with 3 Parity Bits

Parity failures	Error location
$P_2\ P_3$	b_0
$P_1\ P_2\ P_3$	b_1
$P_1\ P_3$	b_2
$P_1\ P_2$	b_3
P_3	P_3
P_2	P_2
P_1	P_1

so the error can be corrected by inverting the value of b_0. Table 1.4 shows the error-correcting actions required on a 4-bit data word with 3 parity bits.

Worked Example 1.6

Detect and correct the error in the odd parity coded word 1110100. The format of the word is $P_3P_2P_1b_3b_2b_1b_0$ where P_3 is the parity of $b_2b_1b_0$, P_2 the parity for $b_3b_1b_0$ and P_1 the parity for $b_3b_2b_1$.

A parity check on the data section of the word gives

Use P_1' for the parity check to distinguish it from P_1 the received parity.

$P_3'\ b_2b_1b_0 = 0100$
$\therefore P_3' = 0$
$P_2'\ b_3b_1b_0 = 1000$
$\therefore P_2' = 1$
$P_1'\ b_3b_2b_1 = 0010$
$\therefore P_1' = 0$

Comparing parity check with parity received

$P_3'P_2'P_1' = 010$
and $P_3P_2P_1 = 111$

Parity failures occur on P_3 and P_1.

The error is therefore located at bit b_2.

See Table 1.4.

Hence by inverting the value of bit b_2, the correct code word is obtained, which is

1110000

Parity Block Checks

An error-correcting parity check on each data word can take a considerable amount of time. It also requires a significant increase in the number of bits needed to transmit a given amount of information. If the occurrence of errors is known to be infrequent, a parity block check system can be used, which will detect and correct one error in a block of data words.

Each data word has a single parity check digit and a parity word is generated at the end of the block. Each bit in the parity word is a check on the corresponding bits in all the data words in the block. Consider a block of 3 words: 101, 100 and 010. Using even parity, the data block becomes

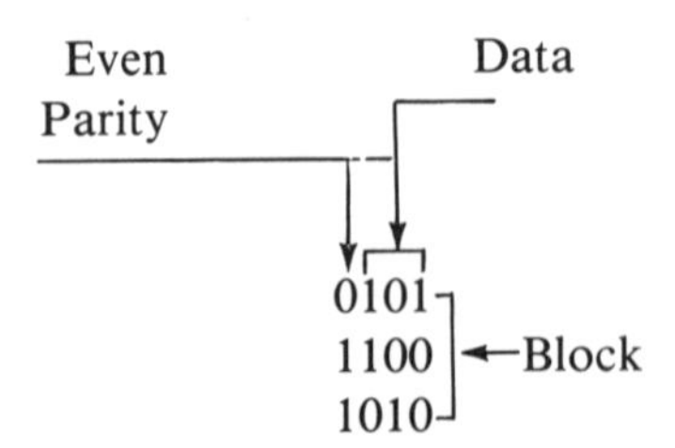

The parity word at the end of the block is obtained by calculating the parity of each column.

```
                        0 | 101
                        0 | 100
                        1 | 010
                          -----
Even parity word →      1   011
```

If an error occurs in a data word, its own parity check will fail, together with one of the bits in the parity word. The error in the data is at the intersection of the row and column containing the failed parity checks.

Worked Example 1.7

The following block of data is received. It has been encoded with odd parity (m.s.b) and the last word is a parity block check. Identify and correct the error in the data. The data is 10000, 01111, 00100, 11100, 11010.

The data block as received, is

```
Word 1              1 | 0000
Word 2              0 | 1111
Word 3              0 | 0100
Word 4              1 | 1100
                      ------
Odd parity block    1   1010
check word          ↑
                    |
          Odd parity bit
```

Generating the parity from the data alone we get

```
 1 | 0000
*1 | 1111
 0 | 0100
 1 | 1100
   ------
 0   1000
       *
```

Remember, the least significant bit is 'bit 0'.

The parity failures indicated by * can be identified and the intersection of the row and column locate the error. The error occurs in bit 1 of word 2 — it should be 0.

The corrected data block is therefore

The parity check data is discarded once the error has been located and corrected. Either use the received parity or recalculate parity from the corrected block.

```
1 | 0000
0 | 1101
0 | 0100
1 | 1100
  ------
0   1010
```

Summary

In this chapter, we have examined a wide range of binary codes that can be used to represent numbers. We, as humans, use the decimal number system, but a binary system is more suited to a machine implementation because two distinct states such as high and low voltage can readily be identified without having to make precise measurements. Methods of converting between binary and decimal number systems have been developed.

Conversion between decimal and pure binary can be quite laborious, as each bit may be dependent on all the decimal digits, and vice versa. BCD overcomes this problem as each decimal digit is determined by the value of a group of 4 bits. There are many BCD codes and weighted versions are generally used. The weights may have negative values. The most common BCD code is 8421 weighted, where each decimal digit is replaced by its value in 4-bit pure binary. BCD codes lead to simple input/output circuits but require complex arithmetic systems.

A class of code that facilitates the measurement of position must only change by 1 bit as the code is incremented or decremented. The most common position sensing code is the Gray code.

Finally, error detection and correction techniques were introduced. This is a vast and complex topic and the subject of much current research. Two relatively simple schemes for error detection were considered. The parity method uses an extra bit per word and the '*m* in *n*' code only allows *m* bits to be set to 1 in an *n* bit word. If the number of parity bits per word is increased, errors may be detected and corrected. Alternatively, a parity word at the end of a block of data enable errors to be located. The concept of error correction is very exciting to the digital engineer. It enables him to correct data errors within digital circuits. Errors can arise when information is transmitted over long distance. Interference is likely to be present and corrupt the data, but error-correcting techniques enable the correct information to be regenerated. In the home, the digitally recorded compact audio discs use error-correcting codes. When a scratch or dirt corrupts the signal, error correction can restore the original signal and thereby remove the annoying clicks that would otherwise appear on the final sound output.

Problems

1.1 Convert the following to binary

(i) 141_{10}
(ii) 0.72265625_{10}
(iii) 21.8125_{10}

1.2 Convert the following to decimal

(i) 111001_2
(ii) 0.001011_2
(iii) 101.101011_2

1.3 How many places are required to represent decimal numbers in the range 1 – 1000 in

(i) binary

(ii) base 8 (octal)
(iii) base 16 (hexadecimal)
(iv) 8421_{BCD}

1.4 Convert the following to base 8

(i) 182_{10}
(ii) 1011110101_2
(iii) 47.75_{10}

1.5 Convert the following to base 16. Use symbols A – F for digit values 10 – 15.

(i) 1011100111000111_2
(ii) 4732_{10}
(iii) 254.03125_{10}

1.6 Convert the following numbers to 8421 BCD

(i) 4723_{10}
(ii) 1010111011_2
(iii) $AA4E_{16}$

1.7 Detect and correct the errors in the following data

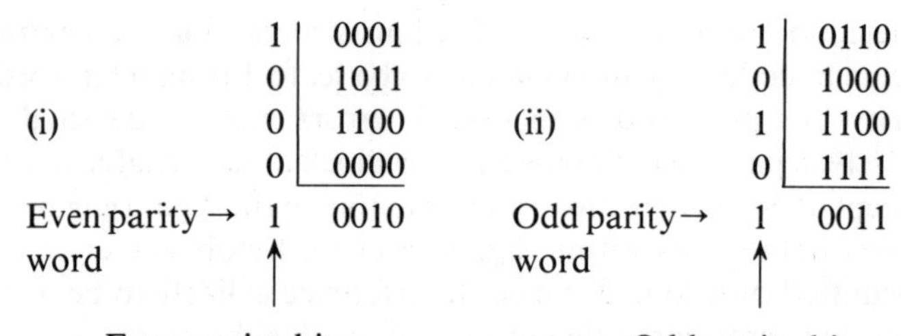

Even parity bit Odd parity bit

1.8 Devise a method whereby two numbers, coded in 8421 BCD can be added together giving the answer in BCD.
1.9 Design an alternative position sensing code to that given in Fig. 1.3.
1.10 Explain the difference between weighted and non-weighted binary coded decimal and give an example of each type of code.

Operations on Binary Data 2

Objectives

- ☐ To define combinational, sequential and storage operations.
- ☐ To relate logic to propositional statements.
- ☐ To define a truth table.
- ☐ To examine binary connectives and thereby define the fundamental logic gates.
- ☐ To apply Boolean algebra to logic design.

We saw in Chapter 1 that information could be encoded into binary, and quantities represented by strings of symbols, where each symbol can only have one of two values. We shall now examine relationships between binary variables.

Electronic logic operations are governed by the same principles and laws as the propositional statements we use in everyday life, as the latter can be regarded as binary functions that are either true or false.

Logical Operations

An electronic circuit that operates on binary data is called a digital logic system. The circuits within the system that carry out the elementary logical operations are called gates. Three general types of logic system can be identified, namely:

(i) combinational logic system,
(ii) sequential logic system,
(iii) storage system.

Any logic system can be represented by a black box, having a set of input lines that receive binary data, and one or more output lines, as shown in Fig. 2.1. If the system is combinational, it has the following property: the output data is a logical function of the value of the input data at the instant the output is measured. If the inputs are represented by the set **I** and the outputs by the set **Z**

then $$\mathbf{Z}_t = f(\mathbf{I}_t) \tag{2.1}$$

where $\mathbf{Z}_t$ is the output and $\mathbf{I}_t$ the input at time t. $f(\mathbf{I}_t)$ represents the function carried out by the logic system. Consequently a given input $\mathbf{I}_t$ will always generate the same

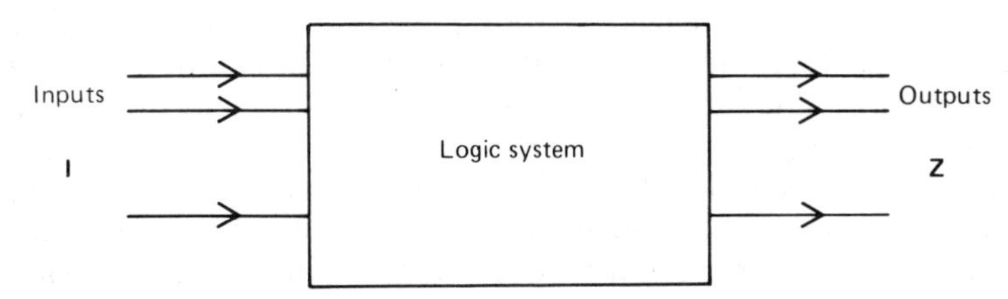

Fig. 2.1 A general logic system.

output from a combinational logic circuit, irrespective of when it is applied to the system.

A given output can be produced by more than one set of inputs. Inputs 4 and 1 would also give output 5.

An adder circuit is an example of a combinational system. If the inputs represent two numbers, say 2 and 3, we would expect an immediate output of 5, and furthermore, the output must be 5, each and every time the inputs are 2 and 3.

A sequential circuit can also be represented by Fig. 2.1 as it also has inputs and outputs. The function of a sequential circuit differs from a combinational logic system in one important respect. The outputs depend not only on the present inputs, but on all the previous inputs over a specific operating time.

Hence for a sequential system

$$\mathbf{Z}_t = f(\mathbf{I}_1 \mathbf{I}_2 \ldots . \mathbf{I}_t) \tag{2.2}$$

where $\mathbf{I}_i$ is the input data at time i.

As the output of the circuit depends on the values of the present and all previous inputs, a given input at time t does not always produce the same output.

Sequential behaviour can be observed in an accumulator circuit that calculates the cumulative sum of all the inputs. Consider the sequence of number 1,4,2,5,2 being input, one at a time, into an accumulator. The outputs are shown in Table 2.1. Each output is the running total of the present and all previous inputs. At time 3, the input is 2 and the resultant output is 7, but at a later time (5) an identical input gives an output of 14. This could not happen in a combinational circuit and provides a means of distinguishing between them. If at any time, a logic system, gives different outputs for the same input, the system must be sequential.

Table 2.1 An Accumulator Operation

Time	Input	Output
1	1	1
2	4	5
3	2	7
4	5	12
5	2	14

In the accumulator circuit, the output is dependent on the magnitude of the previous inputs. In other sequential systems, the behaviour may be dependent on both the magnitude and the order of the input sequence. A telephone handset is an example of the latter. Consider the case where five digits of a six-digit number have already been dialled. The sixth and final digit can have one of only ten values, but the connection is made to one out of a possible maximum of 10^6 subscribers, depending on the value and order of the previous five digits.

The third type of logic system is the storage operation. A storage device has two distinct functions. It can hold information, where data is input and stored but nothing is output, or alternatively information within the memory can be accessed, giving an output without requiring a data input. In memories where more than one number can be stored, additional input information is required to specify the exact location in the memory where the data is to be stored. This input is called the address. Another control input is also necessary to set the memory into its read or write mode. The terminals on a memory circuit are shown in Fig. 2.2.

'Read' means 'access data already stored in memory'.

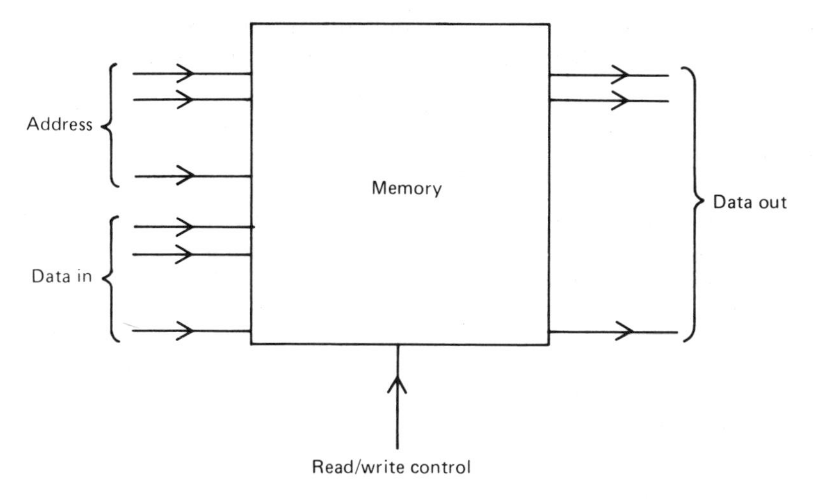

Fig. 2.2 A memory system.

'Write' means 'store data in memory'.

The three types of logic systems are inter-related in a most interesting way. A combinational logic circuit can be converted into a sequential system by applying feedback from some of the outputs to form additional internal inputs. A simple sequential circuit has memory properties and finally a memory circuit can be organised as a combinational logic function. These applications will be examined in the following chapters.

The fundamental electronic circuits common to all types of logic systems are the gates. A logic function is implemented with one or more gates and the relationships between functions are governed by the laws of Boolean algebra.

Logic and Propositional Statements

The formal analysis of binary systems was first investigated by the nineteenth century English mathematician, George Boole, long before the advent of electronics and computers. Boole's logical algebra was developed to test the validity of propositions and verbal statements, but his theorems are directly relevant to the operations which are performed on binary data in electronic systems.

Consider the simple statement about the weather: 'It will snow if the temperature is low and it is cloudy'.

The three variables are 'snow' (the output), 'low temperature' and 'cloudy' (the inputs). Each of these conditions can only be true or false (i.e. true if it is snowing or false if it isn't).

This weather system can be described by a Boolean equation, with the variables S for snow, L for low temperature and C for cloudy, giving

$$S = L \text{ AND } C \tag{2.3}$$

The logic function between the two input variables is AND, which can be identified in the verbal statement. The symbol for AND is . and Equation 2.3 would normally be written

$$S = L.C \tag{2.4}$$

The symbol . may be omitted. S = LC is the same as S = L.C

The equation states that S is true if, and only if, L is true AND C is true. It represents the proposition 'It will snow if the temperature is low and it is cloudy'.

Truth Tables

Every Boolean equation has a truth table that lists the value of the output for each and every possible combination of inputs. If the output is a function of two variables there are 2^2 or 4 possible input combinations. In general, for n variables, the truth table will have 2^n input states.

The truth table for the propositional statement

$$S = L.C$$

The system is assumed to be binary.

The variables can only be true or false.

can be obtained as follows:

The four input values for L and C are (False, False), (False, True), (True, False) and finally (True, True).

The output S can be determined by substituting the input conditions into the statement. S can only be true if L is true and C is true. S is therefore false for all other input conditions and the complete truth table is given in Table 2.2(a). In electronic systems, two voltage levels would be used to represent truth and falsehood and they would be labelled with binary symbols 0 and 1. If truth is represented by logical 1 and falsehood by 0, the binary version of the truth table can be obtained (See Table 2.2b).

Table 2.2 Truth Table for the AND function S = L.C

Output S	Inputs L	C
False	False	False
False	False	True
False	True	False
True	True	True

(a) Truth values

S	L	C
0	0	0
0	0	1
0	1	0
1	1	1

(b) Binary symbols

Worked Example 2.1

Obtain a Boolean equation and its truth table, which specifies the suitability of an applicant for employment with a corporation, requiring employees to be either

(i) unmarried females under 25 years, or
(ii) married males under 25 years, or
(iii) over 25 years.

Let E represent the suitability of the applicant. The input variables are married status M, where M true is married and M false is unmarried, sex S, where S true can represent male and S false, female, and A for age where A true means the applicant is over 25 years.

$\overline{M}$ reads NOT M. When $\overline{M}$ is true, M is false.

Condition 1 is NOT M AND NOT S AND NOT A
Condition 2 is M AND S AND NOT A
Condition 3 is A

Therefore suitability E is ($\overline{M}$ AND $\overline{S}$ AND $\overline{A}$) OR (M AND S AND $\overline{A}$) OR (A)

Replacing AND with the symbol . and OR with +, the Boolean equation is

$$E = (\overline{M}.\overline{S}.\overline{A}) + (M.S.\overline{A}) + A$$

The truth table will contain 8 input conditions as there are 3 input variables. Condition 1 is satisfied by input reference 0, condition 2 by input 6 and condition 3 by inputs 1,3,5, and 7. Hence the truth table is

The input reference number is the decimal value of the inputs where F = 0 and T = 1.

E	M.	S.	A	Input reference
T	F	F	F	0
T	F	F	T	1
F	F	T	F	2
T	F	T	T	3
F	T	F	F	4
T	T	F	T	5
T	T	T	F	6
T	T	T	T	7

Binary Connectives

A binary connective is an elementary function of two input variables. Consider a simple binary system having single output Z and two inputs A and B.

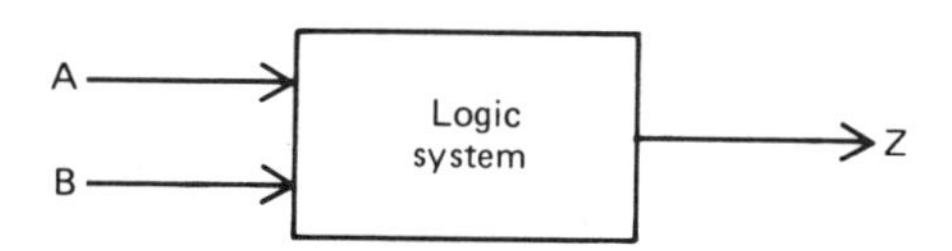

There are four possible combinations of truth values that can occur at the two inputs. They are FF, FT, TF, and TT. As the four possible input states are applied in turn, they will give rise to a sequence of four outputs, each of which can be true or false. Thus a total of 2^4 or 16 different output responses is possible. Each set is a binary connective, and represents a unique function between the input variables and the output. The full list of connectives is given in Table 2.3. It contains all possible functions of two input variables.

F is false. T is true.

Connectives 0 and 15 are independent of the inputs and always give a constant output, whereas 3 and 5 are both independent of one variable, and have outputs identical to A and B, respectively. 10 and 12 are also only dependent on one variable, the outputs taking the opposite value of the inputs. This is the INVERSE or NOT function. The two propositional functions AND and OR occur as connectives 1 and 7, whereas 8 and 14, the inverse of these functions, are called NAND and NOR.

NAND means NOT AND.

Connective 6 is the EXCLUSIVE OR function. It differs from the OR operation in one important respect. The OR function is true if both inputs are true, whereas the EXCLUSIVE OR would be false under these input conditions. Both functions are true when only one input is true.

OR is sometimes known as INCLUSIVE OR.

The logic designer must be aware of the difference between OR and EX.OR. The

Table 2.3 Binary Connectives between Inputs A.B and Output Z

A	Input values F	F	T	T		
B	F	T	F	T	Function	Symbol
0	F	F	F	F	Universal falsehood	0
1	F	F	F	T	AND	A.B
2	F	F	T	F	—	—
3	F	F	T	T	Variable A	A
4	F	T	F	F	—	—
5	F	T	F	T	Variable B	B
6	F	T	T	F	EXCLUSIVE OR	$A \oplus B$
7	F	T	T	T	OR	$A + B$
8	T	F	F	F	NOR	$\overline{A + B}$
9	T	F	F	T	Equivalence	$\overline{A \oplus B}$
10	T	F	T	F	Not B	$\overline{B}$
11	T	F	T	T	—	—
12	T	T	F	F	Not A	$\overline{A}$
13	T	T	F	T	—	—
14	T	T	T	F	NAND	$\overline{A.B}$
15	T	T	T	T	Universal truth	1

distinction is not always obvious in natural language. Consider the following statements:

(i) Safety limit for elevator is twelve persons <u>OR</u> one ton goods.

(ii) We will accept US dollars <u>OR</u> pounds sterling.

OR function can have any number of inputs. EX.OR is a function of only two inputs.

Statement 1 uses the EXCLUSIVE OR connective whereas statement 2 is the OR function.

Logic Gates

A logic gate is a two-state device that implements a binary connective. Most logic devices are electronic, although logic operations can be performed by mechanical, magnetic and even fluid systems. The fundamental gates are AND, OR and NOT and any logical statement can be expressed by using a combination of these three gates.

AND, OR and NOT form a universal set of gates from which all other logic functions can be constructed.

Worked Example 2.2

Express binary connective 13 in terms of AND, OR and NOT gates.

The truth table for binary connective 13 is

A	B	Z
F	F	T
F	T	F
T	F	T
T	T	T

The proposition is:

Output Z is true if A is true and B is true
or A is true and B is false
or A is false and B is false

Expressing this statement as a Boolean equation, we obtain:

$$Z = A.B + A.\overline{B} + \overline{A}.\overline{B}.$$

which reads

Z equals A and B or A and not B or not A and not B

Exercise 2.1

Verify that all the binary connectives dependent on two input variables can be expressed with AND, OR and NOT gates.

Other gates include the EXCLUSIVE OR function

$$Z = A\overline{B} + \overline{A}B$$

and its inversion which is the EQUIVALENCE function

This operation is sometimes called EXCLUSIVE NOR.

$$Z = AB + \overline{A}\overline{B}$$

NAND and NOR gates are the inversion of AND and OR, respectively. The principal logic gates, their symbols, equations and binary truth tables are summarised in Table 2.4.

Boolean Algebra

The principles of logic were developed by George Boole (1815–1884) who, along with Augustus De Morgan, formulated a basic set of rules that govern the relationship between the true – false statements of logic. Boole's work laid the foundation to what later became known as Boolean Algebra. Nearly one hundred years later, Claude Shannon, an American postgraduate at Massachusetts Institute of Technology, realized that Boole's work was relevant in particular to the analysis of switching circuits in telephone exchanges and, more generally, formed a mathematical basis for the electronic processing of binary information.

The theory of electronic logic systems is often called 'switching theory'.

Boolean Theorems

The theorems of Boolean algebra fall into three main groups

Table 2.4 Logic Gates

Name	Symbol	Equation	A	B	Z
AND		$Z = A.B.$	0	0	0
			0	1	0
			1	0	0
			1	1	1
OR		$Z = A + B$	0	0	0
			0	1	1
			1	0	1
			1	1	1
NOT		$Z = \overline{A}$	0		1
			1		0
NAND		$Z = \overline{A.B}$	0	0	1
			0	1	1
			1	0	1
			1	1	0
NOR		$Z = \overline{A + B}$	0	0	1
			0	1	0
			1	0	0
			1	1	0
EXCLUSIVE OR		$Z = A \oplus B$	0	0	0
			0	1	1
			1	0	1
			1	1	0
EQUIVALENCE (EXCLUSIVE NOR)		$Z = \overline{A \oplus B}$	0	0	1
			0	1	0
			1	0	0
			1	1	1

(i) Logical operation on constants.
(ii) Logical operations on one variable.
(iii) Logical operations on two or more variables.

Also known as Huntington's Postulates

Logical Operations on Constants

AND	OR	NOT
$0.0 = 0$	$0+0 = 0$	
$0.1 = 0$	$0+1 = 1$	$\overline{0} = 1$
$1.0 = 0$	$1+0 = 1$	$\overline{1} = 0$
$1.1 = 1$	$1+1 = 1$	

In the AND operations, 0 dominates. If one or more inputs are at 0, then the output becomes 0.

In the OR expressions, a logical 1 on the inputs is dominant. The output becomes 1 if one or more inputs are at 1.

Logical Operations on one variable

AND	OR	NOT
$A.0 = 0$	$A+0 = A$	
$A.1 = A$	$A+1 = 1$	$\overline{\overline{A}} = A$
$A.A = A$	$A+A = A$	
$A.\overline{A} = 0$	$A+\overline{A} = 1$	

Logical Operations on two or more variables

Commutation $A+B = B+A$
$A.B = B.A$

The commutation rule states that there is no significance in the order of the variables. A OR B is identical to B OR A.

Absorption $A+A.B = A$
$A.(A+B) = A$

The absorption rule can be used to simplify logic expressions.

Association $A+(B+C) = (A+B)+C = (A+C)+B = A+B+C$
$A.(B.C) = (A.B).C = (A.C).B = A.B.C.$

The association rule allows variables within the same logic operation to be grouped together in any order. For example, the OR of three variables A, B and C can be achieved either with a three-input OR gate, $A+B+C$, or by using two gates each having two inputs. The first gate can produce the OR of any two input variables and its output is then ORed with the third variable giving $(A+B)+C$ or $(A+C)+B$ or $A+(B+C)$. All four versions are logically identical.

De Morgan's Theorems

$$\overline{A+B} = \overline{A}.\overline{B}$$
$$\overline{A.B} = \overline{A} + \overline{B}$$

$\overline{A + B}$ reads A NOR B.

$\overline{A.B}$ reads A NAND B.

De Morgan's theorems are most important in logic design. They relate AND to NOR and OR to NAND.

Distributive Laws

$$A.(B+C) = A.B + A.C$$
$$A+(B.C) = (A+B).(A+C)$$

Distribution is a process similar to factorization in arithmetic.

Minimization Theorems

$$A.B + A.\overline{B} = A$$
$$(A+B).(A+\overline{B}) = A$$

$$\begin{aligned}(A+B)(A+\overline{B}) &= A.A + A\overline{B} + AB + B\overline{B} \\ &= A + A\overline{B} + AB \\ &= A + A(\overline{B} + B) \\ &= A + A \\ &= A\end{aligned}$$

$A + \overline{A}.B \qquad = A + B$
$A. (\overline{A}+B) \qquad = A.B$

INVERSION is the same as NOT.

In Boolean algebra, as in arithmetic, there is an order of precedence of operations. Bracketed functions are evaluated first, followed by AND, then OR and finally INVERSIONS.

The expression A + B.C means A ORed with the result of B AND C. It is *not* C ANDed with the output of A OR B. The latter would have to be written

$(A+B).\ C$

The digital systems engineer must be aware of the principles of Boolean algebra; however it is not necessary to memorize all the theorems. The operations on logical constants and single variables are rather obvious and can easily be established from first principles and the associative and commutative properties occur in arithmetic. We shall also examine graphical and tabular methods of simplifying logic equations that replace the absorption and minimization theorems. De Morgan's theorems however are important in the design of logic systems. They do not have a parallel in arithmetic and must be learnt.

Verification of Boolean Theorems

Boolean theorems can be verified by logical reasoning or by demonstrating equivalence relationships on a Venn diagram.

Worked Example 2.3. Show that $A + \overline{A} = 1$ using logical reasoning.

Let $F = A+\overline{A}$

The equation reads

'iff' reads 'if and only if'.

F is 1 iff A is 1 or not A is 1

But not A is 1 means A is 0

∴ F is 1 iff A is 1 or A is 0

But A is binary and must be either 0 or 1, so one of the conditions for A is always satisfied.

∴ F is always 1
$\therefore A + \overline{A} = 1$

In a Venn diagram, the binary variables of a function are represented as overlapping areas in a universe. The OR function is the combination or union of areas and the AND function is the intersection or common part between two or more overlapping areas. The NOT function is the remainder of the universe outside a given area. Two functions are equivalent if they define identical areas on a Venn diagram.

Worked Example 2.4. Show that $AB + A\overline{B} = A$ by using Venn Diagrams.

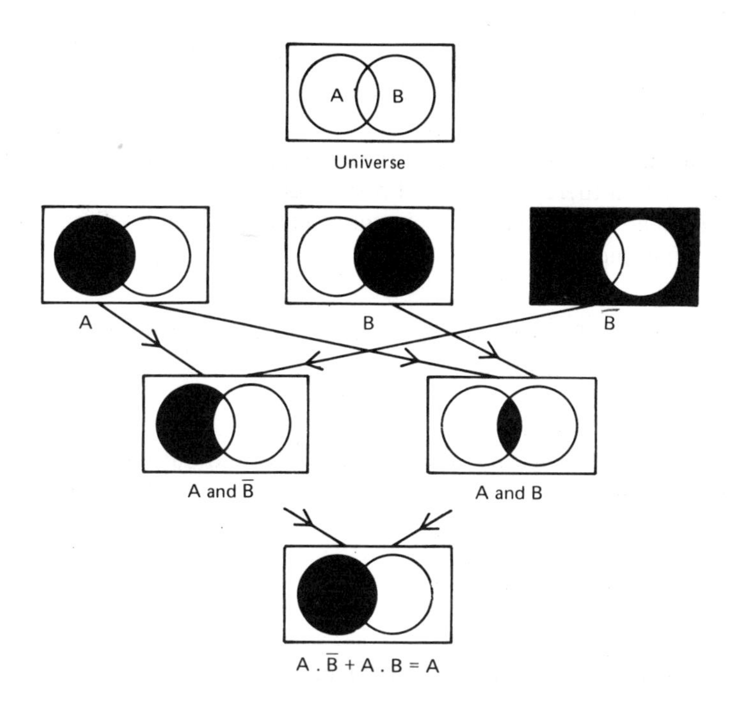

Final Venn diagram is identical to A.

A third method can be used to verify a Boolean identity. The truth tables for both sides of the equation are calculated, and if they are identical for all input values, then the identity is true.

Worked Example 2.5

Verify the De Morgan's theorem $\overline{A.B} = \overline{A}+\overline{B}$ using truth tables.

Given $\overline{A.B} = \overline{A} + \overline{B}$
Let $F_1 = \overline{A.B}$ and $F_2 = \overline{A} + \overline{B}$

Truth table for F_1 and F_2 are

A	B	A.B	$F_1 = \overline{A.B}$
0	0	0	1
0	1	0	1
1	0	0	1
1	1	1	0

and

A	B	$\overline{A}$	$\overline{B}$	$F_2 = \overline{A} + \overline{B}$
0	0	1	1	1
0	1	1	0	1
1	0	0	1	1
1	1	0	0	0

Comparing F_1 and F_2 we see that they are identical for every input AB

Hence $\overline{A.B} = \overline{A} + \overline{B}$

The Principle of Duality

A careful scrutiny of the Boolean theorems will reveal an interesting property. Most theorems occur in pairs and, given one theorem, the other can be obtained by interchanging all the AND and OR gates.

Take, for example, the first absorption rule

$$A + A.B = A$$

If the OR gate is replaced by AND and vice versa, the 2nd absorption theorem is obtained.

$$A.(A+B) = A$$

This natural pairing is an example of the principle of duality, which states that for every logic system there exists a dual form that performs the same function. The dual form can be obtained by interchanging all the AND and OR gates and inverting every input and output.

Worked Example 2.6

Given the function $F = (A+B).C$, obtain its dual form and prove that the dual performs the same function as F.

Given $F = (A+B).C$

The dual may be obtained by inverting all inputs and outputs and interchanging AND and OR gates.

F_D is the dual of F.

Hence $\overline{F}_D = (\overline{A}.\overline{B}) + \overline{C}$

Inverting both sides to obtain F_D gives

F_D reads 'NOT A ANDed with NOT B and then NORed with NOT C'.

$$F_D = \overline{(\overline{A}.\overline{B}) + \overline{C}}$$

F_D is of the form $\overline{X+Y}$ where $X = (\overline{A}.\overline{B})$ and $Y = \overline{C}$

But $\overline{X+Y} = \overline{X}.\overline{Y}$ by De Morgan's theorem

Hence $F_D = \overline{\overline{A}.\overline{B}}.C$

Also $\overline{\overline{A}.\overline{B}} = A+B$ by De Morgan's theorem

$F_D = (A+B).C$

Hence $F_D = F$

Truth Table Equivalence

Truth table equivalence has already been used to verify Boolean theorems. It is however of great practical importance as it provides a quick and simple method of determining equality between complex logic systems.

Boolean algebra provides a mathematical foundation for binary information processing. It can be used to describe complex operations, prove identities and simplify logic systems. In practice, however, Boolean algebra is of limited value to the engineer. The success of a minimization or a proof of an identity depends largely on the person's expertise, and success cannot be guaranteed.

If we take two logic systems where one is considerably more complex than the other, it is important to know whether they perform the same process. If they do, then the engineer would always choose the simpler system. Boolean algebra can be used to establish identities; however, if two systems cannot be proved equivalent by algebra, the result is inconclusive. Either the systems are not equivalent or the engineer's algebra is inadequate.

A better, more reliable method of demonstrating equivalence between two systems is to generate their respective truth tables. If two or more systems have identical truth tables, then they perform the same function.

Worked Example 2.7

If $F_1 = A(A+\overline{B}) + BC(\overline{A}+B) + \overline{B}(A \oplus C)$

and $F_2 = A+C$

determine whether $F_1 = F_2$ by means of truth tables.

F_1 is a function of A B and C. F_2 can be considered a function of A B and C with its output independent of B.

Truth tables are

Remember $A \oplus C$ is the EX.OR function and equal to $A.\bar{C} + \bar{A}.C$

Inputs			Intermediate Terms			Outputs	
A	B	C	$A(A+\bar{B})$	$BC(\bar{A}+B)$	$\bar{B}(A \oplus C)$	F_1	F_2
0	0	0	0	0	0	0	0
0	0	1	0	0	1	1	1
0	1	0	0	0	0	0	0
0	1	1	0	1	0	1	1
1	0	0	1	0	1	1	1
1	0	1	1	0	0	1	1
1	1	0	1	0	0	1	1
1	1	1	1	1	0	1	1

If the output cannot be calculated directly from the Boolean equations and the input values, evaluate any necessary intermediate terms first.

By comparing the output columns, we can see that F_1 and F_2 always have the same value for each and every output.

Hence $F_1 = F_2$

Exercise 2.2

Use Boolean algebra to determine whether or not the following functions are equivalent.

$F_1 = \overline{(A+B).(A+C)} + \overline{A+B+C}$

and $F_2 = \bar{A}\,\bar{B}\,\bar{C} + ((A\bar{C}) \oplus \bar{B})$

Confirm your finding by means of truth tables.

Summary

Logic operations can be divided into three broad and inter-related groups. In a combinational operation, the output at a given time is a function of the inputs at that instant. In a sequential operation, the output is a function of the value and order of a series of inputs over a given period of time. A storage operation can hold input data and output it at a later time. Logic functions are similar to propositional statements and can be completely and uniquely specified by a truth table that gives the output values for each and every possible combination of inputs.

A binary connective is a function that relates the output of a binary system to its inputs. By considering all possible functions of two inputs the fundamental logic operations can be identified.

Finally, the relevance of Boolean algebra to logic design was assessed. Boole's theorems, which can be verified by logical reasoning, using Venn diagrams or truth table equivalence, are directly applicable to digital logic systems. However, in practice, truth table equivalence may provide a quicker and more efficient means of proving an identity than the use of Boolean algebra.

Problems

2.1 Identify the following 2-input functions

(i) The output is 0 only when both inputs are 1.
(ii) The output is 1 provided the inputs are different.
(iii) The output is 1 provided no more than 1 input variable is 1.
(iv) The only time the output is 0 is when both inputs are 0.

2.2 Prove the following relationships using Venn diagrams

(i) $A + \overline{A}B = A + B$
(ii) $\overline{A.B} = \overline{A} + \overline{B}$

2.3 Devise truth tables for

(i) a three-input NAND gate.
(ii) a three-input NOR gate.

2.4 Devise a truth table for a logic system, with two inputs A and B, where the output F is equal to B if A is 0, and $\overline{B}$, if A is 1. What gate is required to implement this function?

2.5 Obtain truth tables for the following equations

(i) $F = \overline{A} + B.\overline{C} + \overline{A.B}$
(ii) $F = (A + B).(\overline{A} + \overline{B} + C)$
(iii) $F = \overline{(A.B) + \overline{\overline{A} + \overline{B}}}$

2.6 Prove the following identities using Boolean algebra

(i) $\overline{(A + B).(\overline{\overline{A}.\overline{B}})} = \overline{A}.\overline{B}$
(ii) $(A + \overline{B}).(\overline{A} + \overline{B} + C) = A.C + \overline{B}$
(iii) $(\overline{A \oplus B}) + (B \oplus C) + (A \oplus C) = 1$

2.7 Verify the identities in Problem 2.6 by using truth tables.

2.8 (i) Show that $\overline{A.B}$ is the dual of $\overline{A} + \overline{B}$
(ii) What is the dual of the following circuit?

$$F = \overline{A}.\overline{B} + \overline{A + C}$$

2.9 Convert the equation $F = A \oplus (B \oplus C)$ into a form that requires only AND, OR and NOT gates and obtain its truth table.

2.10 A company is controlled by managing director A, financial director B and two elected members of the board C and D. A needs the support of one other and B needs the support of two others in order to make a decision. Obtain a truth table for the decision-making strategies and identify the voting when the decision goes

(i) against A.
(ii) against B.

Combinational Logic Design 3

Objectives

- ☐ To define logic levels in an electronic circuit.
- ☐ To investigate canonical logic forms.
- ☐ To examine the Karnaugh map method of logic minimization.
- ☐ To design NAND logic systems and NOR logic systems.
- ☐ To define and utilize 'don't care' conditions.
- ☐ To investigate electronic 'hazards'.
- ☐ To present a tabular method of logic minimization that is suitable for computer implementation.
- ☐ To examine the principles of cellular logic.

This chapter introduces the principles of combinational logic design at the gate level. We start with the definition of logic levels and proceed with the design and optimization of combinational logic functions that can be built out of discrete gates. The final section of the chapter deals with cellular logic, where design principles are required to produce an individual cell, which can then be repeated, using integrated circuit methods, to produce complex devices.

Assignment of Logic Levels

Before designing a logic system, a method of representing binary data within an electronic system must be defined. Each input and output can only have two states, which represent the binary symbols 0 and 1. Consider an electronic circuit having two inputs, A and B, and a single output, F. An example of input/output behaviour is given in Table 3.1 where the allowable voltage states are 0 V and 5 V.

Table 3.1 Electrical Behaviour of a Binary System, Showing the Voltages at the Inputs and Outputs

Inputs A	Inputs B	Output F
0V	0V	0V
0V	5V	0V
5V	0V	0V
5V	5V	5V

If positive coding is used where the symbol 1 is assigned to the positive or higher voltage and 0 to the negative or lower voltage, the truth table for the circuit becomes

A	B	F
0	0	0
0	1	0
1	0	0
1	1	1

and it performs the AND function F = A.B.

Using negative coding where 0 V ≡ 1 and 5 V ≡ 0 the truth table now becomes

A	B	F
0	0	0
0	1	1
1	0	1
1	1	1

NAND is the inverse of AND

and the same electronic system now performs the OR function F = A + B. This is an important and unexpected result. Inverting the binary state assignments of an electronic system does *not* result in the inversion of the circuit's function. A circuit that with positive logic coding, performs the AND function, will become an OR gate with negative logic coding. It does *not* become a NAND gate. The implications of this behaviour are far-reaching. Manufacturers must specify, and users adhere to the logic coding of digital devices. If the coding is changed, a logic system will perform a completely different function, although its circuit and electronic behaviour has not been altered. Most manufacturers use positive coding, although negative, and in some cases, mixed coding, can be used.

Exercise 3.1 An electronic circuit has three inputs, A,B and C, and one output, F, all of which are restricted to either −5 V or +15 V. Its electrical behaviour is

A	B	C	F
−5V	−5V	−5V	−5V
−5V	−5V	+15V	+15V
−5V	+15V	−5V	−5V
−5V	+15V	+15V	−5V
+15V	−5V	−5V	−5V
+15V	−5V	+15V	−5V
+15V	+15V	−5V	+15V
+15V	+15V	+15V	+15V

What logic function does the circuit perform with
(a) positive coding
(b) negative coding?

Specification of a Combinational Logic Circuit

In order to design a combinational circuit the inputs and their corresponding outputs, must be identified. A combinational system having more than one output, can be considered as several single output logic circuits operating on common inputs. Once the input/output relationship has been defined, a truth table can be calculated.

Consider a combinational logic system that accepts 3-bit binary numbers and indicates if they are in the range 3_{10} to 5_{10}. Three input lines A, B and C are required together with a single bit output F, where F = 1 means the input is within range, otherwise F = 0.

The truth table for the system can now be compiled. There are eight possible input values. For each input, the designer must determine whether or not the numerical value of ABC is within the specified range. The truth table for the logic system is given in Table 3.2.

Table 3.2 Truth Table for 3 to 5 Range Indicator

A	B	C	F
0	0	0	0
0	0	1	0
0	1	0	0
0	1	1	1
1	0	0	1
1	0	1	1
1	1	0	0
1	1	1	0

The 1st Canonical Form

The 1st canonical form of a combinational logic function can be obtained from the truth table. It consists of a set of minterms that are AND functions of the input variables or their inversions. The outputs of the AND operations are ORed together to give a single output. In a 1st canonical form circuit the ANDing is always carried out on the input data before the OR operation.

Alternative name for 1st canonical form is Sum of Products or SOP.

The 1st canonical form can be obtained from a propositional description of a truth table. Taking the truth table 3.2 as an example, we have

F is 1 iff A is 0 and B is 1 and C is 1
or A is 1 and B is 0 and C is 0
or A is 1 and B is 0 and C is 1

which can be expressed as the Boolean equation:

$$F = \overline{A}BC + A\overline{B}\,\overline{C} + A\overline{B}C \qquad (3.1)$$

whose circuit diagram is given in Fig. 3.1.

The rule for obtaining the 1st canonical form from a truth table is as follows:

For every entry with an output at 1, create an AND of the input variables. The variable is used if its value is 1, or its inverse, if 0. The outputs of the AND gates are then ORed together to obtain the final output.

If input A is 1, use A. If input A is 0, use $\overline{A}$.

Shorthand Notation for the 1st Canonical Form

In order to avoid having to write down long Boolean equations for complex systems, the following shorthand notation will be adopted.

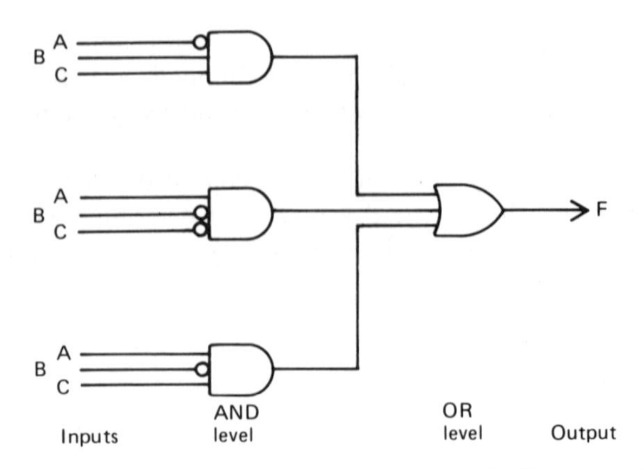

Fig. 3.1 A 1st canonical form circuit (—○ on inputs indicates an inversion).

For a 1st canonical form function

$$f(\text{ABC}) = \overline{\text{A}}\text{BC} + \text{A}\overline{\text{B}}\overline{\text{C}} + \text{A}\overline{\text{B}}\text{C}$$

the variables in the minterms are replaced with their binary values, giving

$$f(\text{ABC}) = 011 + 100 + 101$$

Each minterm is regarded as a number and its decimal value obtained

$$f(\text{ABC}) = 3 + 4 + 5$$

The function is now represented by the set of the decimal numbers

Hence $f(\text{ABC}) = \Sigma(3, 4, 5)$

where Σ indicates the 1st canonical form.

Worked Example 3.1 Express $\text{F(ABCD)} = \Sigma(3, 4, 9, 10)$ as a Boolean equation.

Minterm 3 is 0011 and represents $\overline{\text{A}}\,\overline{\text{B}}\,\text{C}\,\text{D}$
Minterm 4 is 0100 and represents $\overline{\text{A}}\,\text{B}\,\overline{\text{C}}\,\overline{\text{D}}$
Minterm 9 is 1001 and represents $\text{A}\,\overline{\text{B}}\,\overline{\text{C}}\,\text{D}$
Minterm 10 is 1010 and represents $\text{A}\,\overline{\text{B}}\,\text{C}\,\overline{\text{D}}$
Hence $\text{F} = \overline{\text{A}}\,\overline{\text{B}}\,\text{C}\,\text{D} + \overline{\text{A}}\,\text{B}\,\overline{\text{C}}\,\overline{\text{D}} + \text{A}\,\overline{\text{B}}\,\overline{\text{C}}\,\text{D} + \text{A}\,\overline{\text{B}}\,\text{C}\,\overline{\text{D}}$

$\text{F} = \Sigma(3,4,9,10)$ could be a function of more than 4 variables. If it is, the number must be specified with the equation.

If the number of variables in the function is not specified, it can be assumed to be the index of the first integer power of 2 greater than the highest decimal assignment in the function set. In Worked Example 3.1, the highest decimal assignment is 10. The first integer power of 2 greater than 10 is $2^4 = 16$. The index is 4, hence there are 4 variables in the function.

The 2nd Canonical Form

Alternative name is Product of Sums or POS.

The 2nd canonical form is an alternative hardware structure to the first form. The input variables are ORed together to form maxterms which are then ANDed together to give the final output. The OR operation is always carried out before the AND.

Important. Compare with 1st canonical form structure.

The 2nd canonical form can be obtained from the truth table by defining the function by its F = 0 terms. Taking truth table 3.2 we have

F is 0 iff A is 0 and B is 0 and C is 0
or A is 0 and B is 0 and C is 1
or A is 0 and B is 1 and C is 0
or A is 1 and B is 1 and C is 0
or A is 1 and B is 1 and C is 1

which gives the Boolean equation

$$\overline{F} = \overline{A}\,\overline{B}\,\overline{C} + \overline{A}\,\overline{B}C + \overline{A}B\overline{C} + AB\overline{C} + ABC \qquad (3.2)$$

Note $\overline{F}$ is F = 0

Inverting both sides gives

$$F = \overline{\overline{A}\,\overline{B}\,\overline{C} + \overline{A}\,\overline{B}C + \overline{A}B\overline{C} + AB\overline{C} + ABC} \qquad (3.3)$$

Equation 3.3 is of the form $F = \overline{X_1 + X_2 + \ldots X_n} = \overline{X}_1.\overline{X}_2.\overline{X}_n$

Applying De Morgan's theorem to RHS gives

$$F = \overline{\overline{A}\,\overline{B}\,\overline{C}}\;.\;\overline{\overline{A}\,\overline{B}C}\;.\;\overline{\overline{A}B\overline{C}}\;.\;\overline{AB\overline{C}}\;.\;\overline{ABC}. \qquad (3.4)$$

Applying the second De Morgan theorem to each NAND term in Equation (3.4) gives

$$F = (A+B+C).(A+B+\overline{C}).(A+\overline{B}+C).(\overline{A}+\overline{B}+C).(\overline{A}+\overline{B}+\overline{C}) \qquad (3.5)$$

Note the 'OR followed by AND' structure.

This is the 2nd canonical form of the logic and its circuit diagram is given in Fig. 3.2.

The 2nd canonical form can be obtained directly from the truth table without having to resort to lengthy algebra. Comparing Equation 3.5 with truth table 3.2 we can identify each maxterm with an entry that gives a zero output, provided we represent each inverted variable in the maxterm with a 1 and each normal variable with a 0.

A	B	C	F
0	0	0	0
0	0	1	0
0	1	0	0
0	1	1	1
1	0	0	1
1	0	1	1
1	1	0	0
1	1	1	0

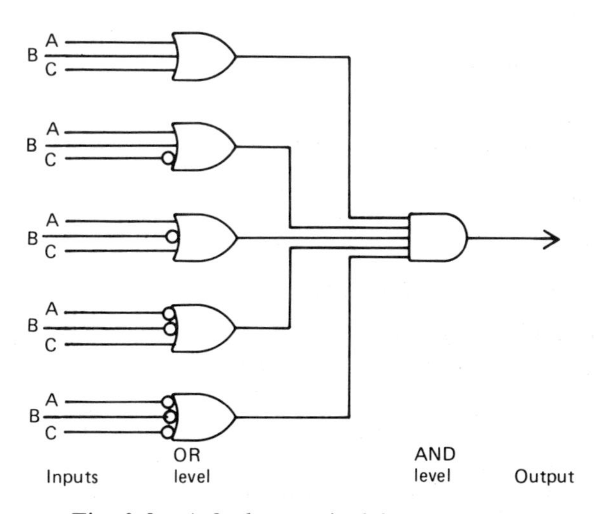

Fig. 3.2 A 2nd canonical form circuit.

For example: Input ABC = 000 gives F=0 and forms the maxterm $(A+B+C)$.
Input ABC = 001 gives F=0 and forms the maxterm $(A+B+\overline{C})$ and so on.

Compare this rule with the 1st canonical rule. It is the opposite in every respect.

The rule for obtaining the 2nd canonical form is therefore: for each entry in the truth table with an output at 0, create an OR of the inputs using the normal variable, if its value is 0, or its inverted value, if the input is at 1. The output of the OR gates are ANDed together to give the final output.

Shorthand Notation for the 2nd Canonical Form

Compare with 1st canonical form, where the normal variable implied 1 and the inverted variable implied 0.

The 2nd form can be represented as a set of numbers identifying the maxterms. Normal variables are assigned 0 and inverted variables 1.

$$\text{Thus } F = (A+B+C).(A+B+\overline{C}).(A+\overline{B}+C).(\overline{A}+\overline{B}+C).(\overline{A}+\overline{B}+\overline{C})$$

gives binary values

	(000)	(001)	(010)	(110)	(111)
and decimal values	0	1	2	6	7

Hence $F = \Pi\ (0,1,2,6,7)$
where Π indicates the 2nd canonical form.

If the 1st and 2nd canonical circuits are obtained from the same truth table, they may be quite different from each other, but will always produce identical outputs for the same inputs. They both perform identical operations on the input data.

Worked Example 3.2 Obtain the 1st and 2nd canonical forms for the EXCLUSIVE OR function.

The truth table for EX.OR is

A	B	F
0	0	0
0	1	1
1	0	1
1	1	0

1st canonical form minterms are:

F = 1 when A is 0 and B is 1, giving $\overline{A}B$
or A is 1 and B is 0, giving $A\overline{B}$

ORing the minterms gives

$$F = \overline{A}B + A\overline{B}$$

For the 2nd canonical form maxterms

F = 0 when A is 0 and B is 0, giving $(A+B)$
also A is 1 and B is 1, giving $(\overline{A}+\overline{B})$

ANDing the maxterms gives

$$F = (A+B).(\overline{A}+\overline{B})$$

[Note also $\overline{A}B + A\overline{B} = (A+B)(\overline{A}+\overline{B})$]

Conversion Between Canonical forms

If the set representation is used, logic functions can readily be converted between 1st and 2nd canonical forms. The set representing the minterms must contain all inputs except for those that are maxterms and vice versa. So for a 3-variabled function, the set of all possible inputs is

$$\mathbf{I} = (0,1,2,3,4,5,6,7,) \tag{3.6}$$

If the minterm set is **m**, then the maxterm set **M** is

$$\mathbf{M} = \mathbf{I} - \mathbf{m} \tag{3.7}$$

Therefore if $\mathbf{m} = (0,5,6,7)$ then **M** would be (1,2,3,4), so the 1st canonical form is

$$F = \Sigma\ (0,5,6,7)$$

and its corresponding 2nd canonical form is

$$F = \Pi\ (1,2,3,4)$$

Worked Example 3.3

If the 1st canonical form of a logic function is

$$F = \overline{A}\,\overline{B}\,\overline{C} + \overline{A}\,\overline{B}C + A\overline{B}C + ABC$$

What is the equation of the 2nd canonical form?

Given $F = \Sigma\ (0,1,5,7)$

Hence $\mathbf{m} = (0,1,5,7)$

But F is a function of three variables

$$\therefore \quad \mathbf{I} = (0,1,2,3,4,5,6,7)$$

$$\text{So } \mathbf{M} = \mathbf{I} - \mathbf{m}$$

$$= (0,1,2,3,4,5,6,7) - (0,1,5,7)$$

$$= (2,3,4,6)$$

$$\therefore \quad F = \Pi\ (2,3,4,6)$$

$$= (A+\overline{B}+C).(A+\overline{B}+\overline{C}).(\overline{A}+B+C).(\overline{A}+\overline{B}+C)$$

Remember. In the 2nd canonical form $A \rightarrow 0$ and $\overline{A} \rightarrow 1$

Minimal Canonical Forms

The canonical forms obtained from the truth table can, in most cases, be simplified or minimized while the AND/OR structure of the 1st form or the OR/AND of the 2nd form are still maintained. The minimization theorems from Boolean algebra could be used directly on the logic equations, but this is a somewhat open-ended exercise, and it is quite often difficult to tell when the simplest version has been obtained.

A graphical method of minimizing logic functions was devised by Karnaugh in 1953. It is based on the minimization theorems and Venn diagrams but also guarantees that the circuit is in its simplest form.

The Karnaugh Map

Also known as a K-map.

Consider a Venn diagram containing the binary variable A, occupying half the universe.

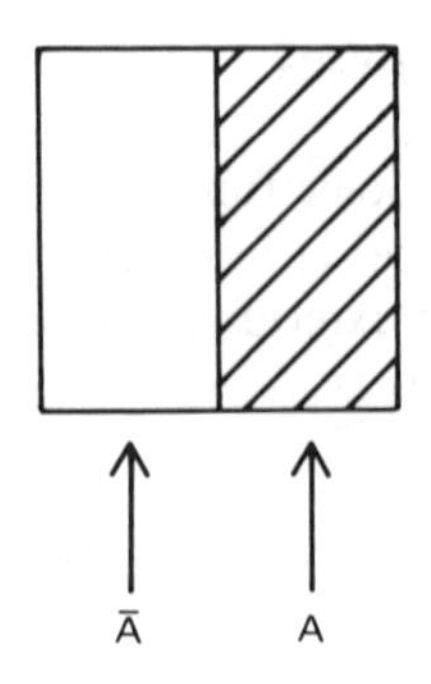

The areas must overlap as A and B are not disjoint.

Now, if a second variable B is introduced, which partitions the universe horizontally, we get

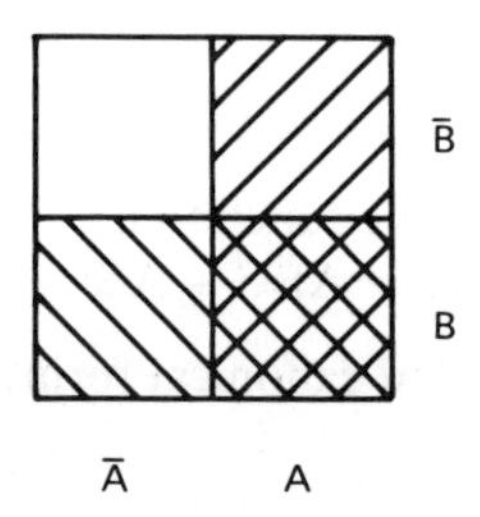

Replacing the shaded areas with labels on the axes, we obtain a 4-celled K-map, and each cell can contain the output of a logic function when its inputs have the values of the cell co-ordinates. A 4-celled K-map can represent any 2-variabled logic function. In the K-map of an AND function, a 1 is placed in the cell corresponding to the intersection of areas A and B on the Venn diagram, whereas the OR function is the union of these areas (see Fig. 3.3).

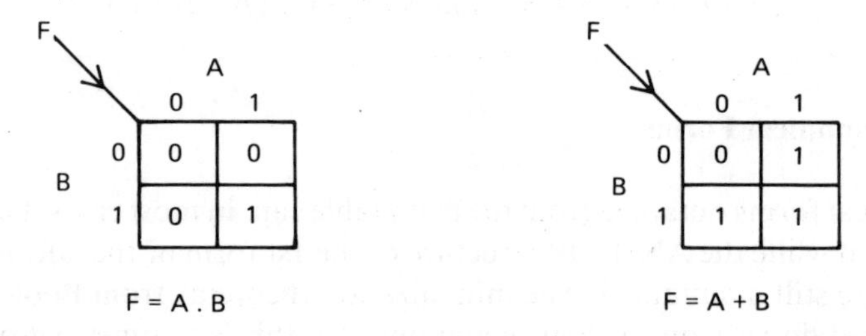

Fig. 3.3 Karnaugh maps for the AND and OR functions.

Karnaugh maps for 3- and 4-variabled functions can be drawn in two dimensions. Three - dimensional maps can represent functions of up to 6 variables. As the areas representing the individual variables must overlap, each axis must be labelled in Gray code and cannot be extended to more than two variables. The Gray code labelling is 00, 01, 11, 10 for adjacent rows or columns.

This is not pure binary.

Maps for 3- and 4-variabled functions are shown in Fig. 3.4. Each cell corresponds to one input state on the truth table and can be labelled with the appropriate output. The address of each cell is the value of its co-ordinates, expressed as

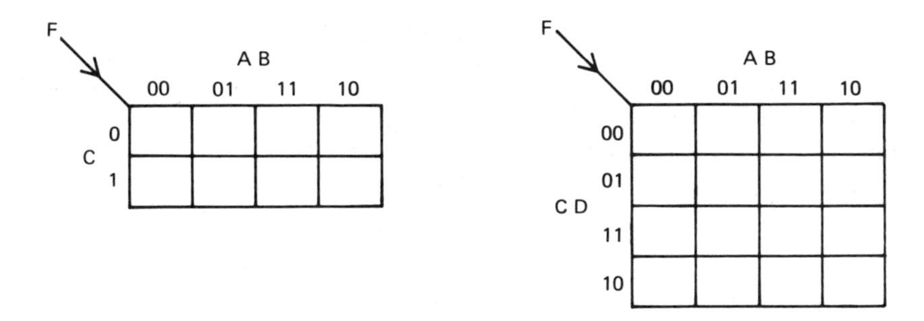

Fig. 3.4 Karnaugh maps for 3- and 4-variabled functions.

a number and is identical to the decimal assignments used in the set notations for the canonical forms.

Worked Example 3.4

Obtain K-maps for the functions:
(a) F = Σ (0,2,4,9,11)
(b) F = Π (3,5,7)

(a) F = Σ (0,2,4,9,11) is a 1st canonical function of 4 variables, say A,B,C and D. For each minterm a 1 is entered into the K-map, giving

F: CD \ AB	00	01	11	10
00	1_0	1_4	0_{12}	0_8
01	0_1	0_5	0_{13}	1_9
11	0_3	0_7	0_{15}	1_{11}
10	1_2	0_6	0_{14}	0_{10}

The address of each cell has been given here as a subscript. Note carefully the order of the numbering due to the Gray coding of the axes.

All remaining cells are set to 0.
(b) For F = Π (3,5,7) we have a 2nd canonical form function of 3 variables. The maxterm set defines the input conditions where the output is 0. Therefore 0s are entered into the K-map locations defined by the maxterm set and the remaining cells are set to 1.

Π means 2nd canonical form. Remember F = 0 for a maxterm.

The K-map is therefore

F: C \ AB	00	01	11	10
0	1_0	1_2	1_6	1_4
1	1_1	0_3	0_7	0_5

Note the cell addresses.

Minimization of canonical forms using Karnaugh Maps

A K-map of a logic function can be used to obtain the simplest version of either the 1st or 2nd canonical form.

Consider the 3-variabled logic function $F = \Sigma\,(1,2,5,6)$.

Its K-map is

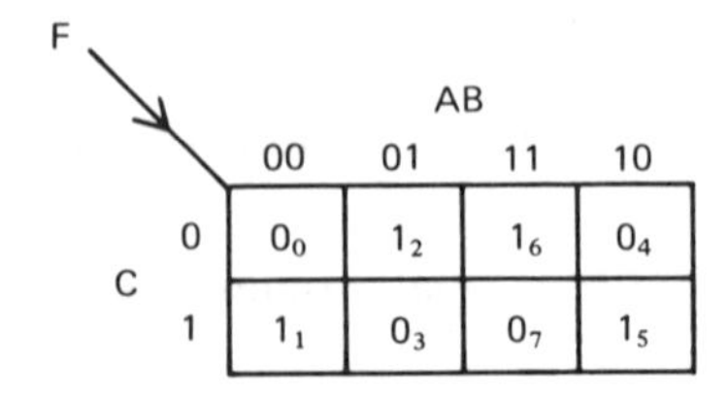

and full equation

$$F = \overline{A}\,\overline{B}C + \overline{A}B\overline{C} + A\overline{B}C + AB\overline{C} \qquad (3.8)$$

Applying the minimization theorem $XY + X\overline{Y} = X$ to minterms 2 and 6, where $X = B\overline{C}$ and $Y = A$, terms $\overline{A}B\overline{C} + AB\overline{C}$ become $B\overline{C}$, which is a considerable simplification.

The minimization theorem is applicable to any minterms on the K-map that occupy adjacent cells. However by using the map alone, the variables that can be eliminated between a pair of minterms, can be detected, without having to resort to algebra.

The procedure is as follows. Neighbouring cells set to 1 are looped together as in Fig. 3.5. K-maps are considered to be continuous at their edges so minterms 1 and 5 are also adjacent and form loop 2.

The function can now be described as being 'F = 1 if the inputs are in loop 1 or loop 2'. By examining the inputs that define loop 1 we see that B must be 1 and C must be 0, but it does not matter what value A takes, since we still remain within the loop if A is 0 or 1. Loop 1 is therefore independent of A and has the logical function $B.\overline{C}$. Similarly for loop 2, B must be 0 and C must be 1, but A can again be 0 or 1. Loop 2 is therefore $\overline{B}.C$ and the proposition F = 1, if in loop 1 or loop 2, may be written as the equation

This is much simpler than Equation 3.8. However, it still retains the AND/OR structure of a 1st canonical form function.

$$F = B\overline{C} + \overline{B}C \qquad (3.9)$$

A similar technique can be used to minimize the 2nd canonical form by looping adjacent 0s on the K-map.

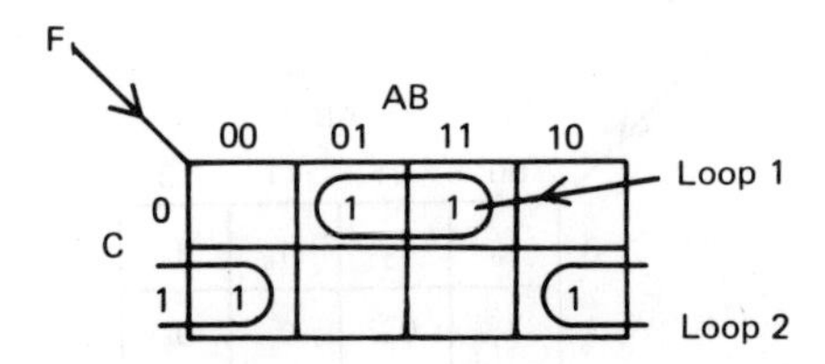

Fig. 3.5 Looping on a 3-variable K-map (blank cells are assumed to be zero).

The 2nd canonical form of Equation 3.8 is

$$F = \Pi\ (0,3,4,7)$$

and its K-map is

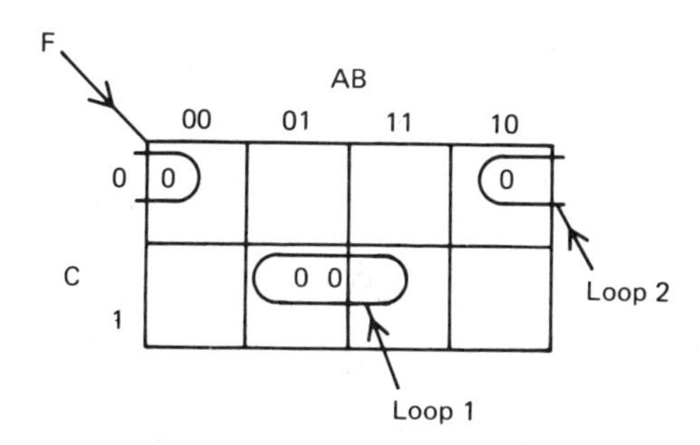

Blank cells are assumed to be set to 1 — omitted for clarity.

Consider maxterms 3 and 7 which are

$$(A+\overline{B}+\overline{C}).\ (\overline{A}+\overline{B}+\overline{C})$$

This expression is of the form $(X+Y)\ (X+\overline{Y}) = X$ where $X = (\overline{B}+\overline{C})$ and $Y = A$. The maxterms can therefore be simplified to

See minimization theorems.

$$(\overline{B}+\overline{C})$$

By looping adjacant cells at 0, an input variable can be eliminated. In loop 1, B must be 1 and C must be 1 but A can be either 0 or 1. Loop 1 is therefore independent of A and the simplified maxterm is $(\overline{B}+\overline{C})$.

Remember. In the 2nd canonical form $1 \rightarrow \overline{B}$ and $0 \rightarrow B$.

The minimal 2nd canonical form for the function is

$$F = (\overline{B}+\overline{C}).\ (B+C) \qquad (3.10)$$

Rules for Looping on a Karnaugh Map

1. Loops must contain 2^n adjacent cells set to 1 (or 0 for the 2nd canonical form). A single cell (loop of 2^0) cannot be simplified. A loop of 2 (2^1) is independent of 1 variable. A loop of 4 (2^2) is independent of 2 variables, and in general a loop of 2^n is independent of n variables. Therefore to obtain the simplest functions, use the largest possible loops.
2. All cells set to 1 must be covered when specifying the minimal form of the function (0 for the 2nd canonical form).
3. Loops may overlap provided they contain at least one otherwise unlooped cell.
4. Any loop that has all its cells included in other loops, is redundant.
5. Loops must be square or rectangular. Diagonal or L-shaped loops are invalid.
6. There may be different ways of looping a K-map as there is not necessarily a unique minimal form for a given circuit.
7. The edges of a K-map are considered to be adjacent. A loop can leave the top of a K-map and re-enter the bottom and similarly for the sides.

Worked Example 3.5

Find the minimal 1st and 2nd canonical forms of the function

$$F = \Sigma\ (3,4,5,6,7,8,10,12,14)$$

The function has four variables and its K-map is

F

CD \ AB	00	01	11	10
00	0_0	1_4	1_{12}	1_8
01	0_1	1_5	0_{13}	0_9
11	1_3	1_7	0_{15}	0_{11}
10	0_2	1_6	1_{14}	1_{10}

Cells (4,5,6, and 7) form a loop of 4 (the largest possible loop for this function). This is loop 1. Cells (8,10,12,14) form another loop of 4 (loop 2) and cells (4,6,12,14) also appear to be a third loop of 4 (loop 3) but all the individual cells are included in either loop 1 or loop 2. Loop 3 is therefore redundant. The final cell to be covered is 3, which combines with 7 to give a loop of 2, (loop 4). The looped K-map is therefore

For clarity when generating a 1st canonical form function from a K-map you may omit cell addresses, and cells at 0 can be left blank.

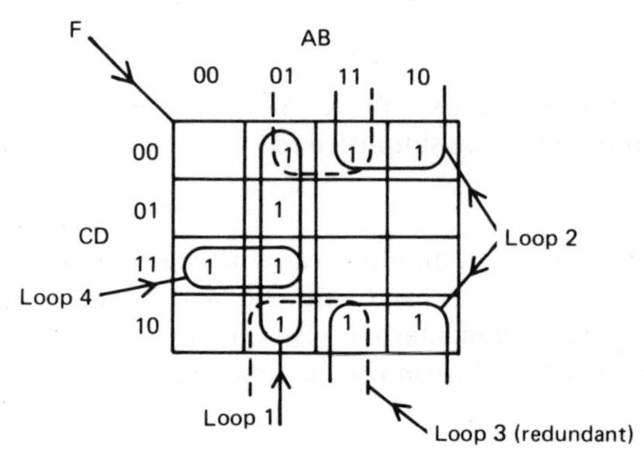

Loop 1 requires A = 0 and B = 1 but is independent of C and D.
Loop 1 is therefore $\overline{A}.B$.
Loop 2 requires A = 1 and D = 0, but is independent of B and C, giving $A\overline{D}$.
Loop 3 is redundant and can be ignored and finally loop 4 has A = 0, C = 1 and D = 1 but is independent of B, giving $\overline{A}CD$.

The minimized version of the 1st canonical form is therefore

$$F = \overline{A}B + A\overline{D} + \overline{A}CD$$

The K-map itself remains the same.

The minimal 2nd canonical form is obtained by looping the zeros on the K-map, giving

Blank cells are assumed to be F = 1 and are omitted for clarity.

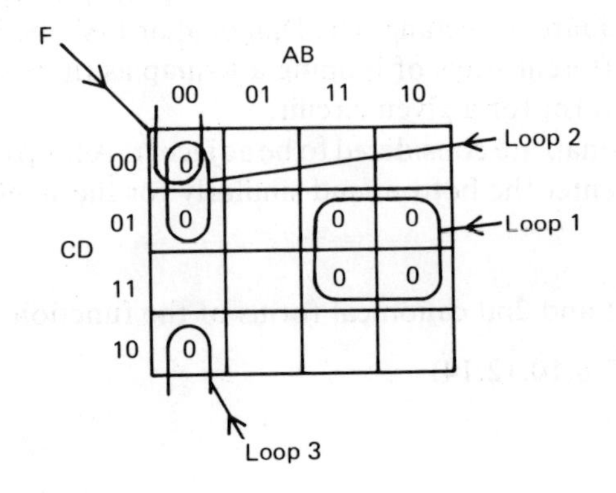

Loop 1 represents $(\overline{A} + \overline{D})$, loop 2, $(A + B + C)$ and loop 3, $(A + B + D)$ giving

$$F = (\overline{A} + \overline{D}).(A + B + C).(A + B + D)$$

NAND Logic

AND, OR and NOT form a universal set of logic gates from which all logic systems can be constructed. It can however be shown that these three functions may be carried out with NAND circuits.

A NAND gate will perform the NOT operation if its inputs are common.

Let $F = \overline{A.B}$
If $A = B$
Then $F = \overline{A.A}$
But $A.A = A$
$\therefore F = \overline{A}$

Hence A F be replaced by A F

The AND function can be replaced by a NAND gate with its output inverted by a second NAND gate because

$$A.B = \overline{\overline{A.B}}$$

Therefore A B F is equivalent to A B F

De Morgan's theorem states that

$$\overline{A.B} = \overline{A} + \overline{B}$$

Inverting all As and Bs gives

$$A + B = \overline{\overline{A}.\overline{B}}$$

This OR function can be implemented with three NAND gates. If the inputs are inverted and then NANDed, the OR function is obtained.

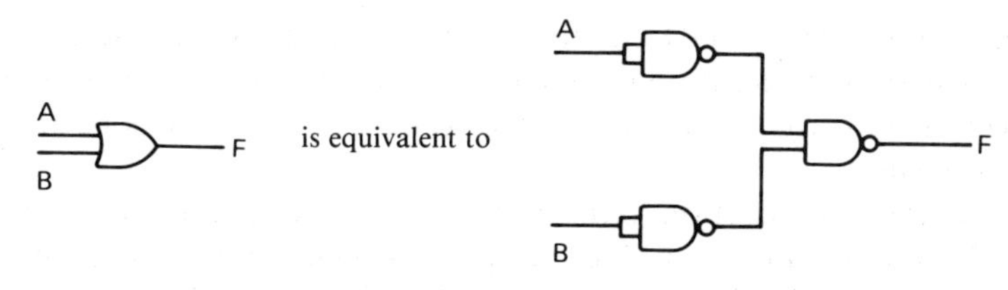

NAND is therefore a universal gate — any logic circuit can be constructed solely out of NAND gates.

NAND logic is particularly relevant to 1st canonical form logic systems. Consider the function

$$F = A\overline{B} + BC \qquad (3.11)$$

—O on an input line means the variable is inverted.

The circuit is

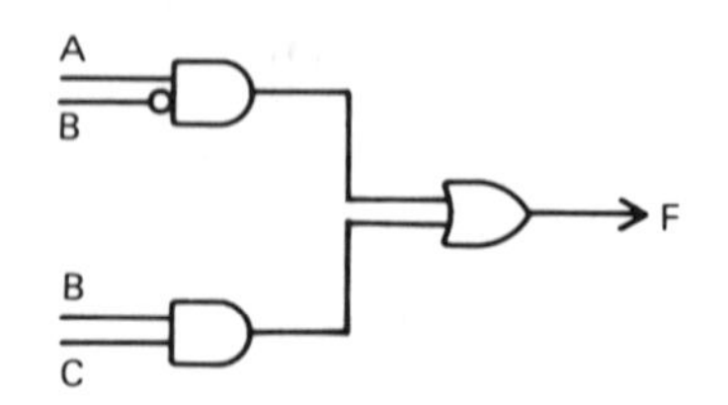

The AND and OR gates can be replaced with their NAND equivalents giving

In complex circuits use gate reference numbers.

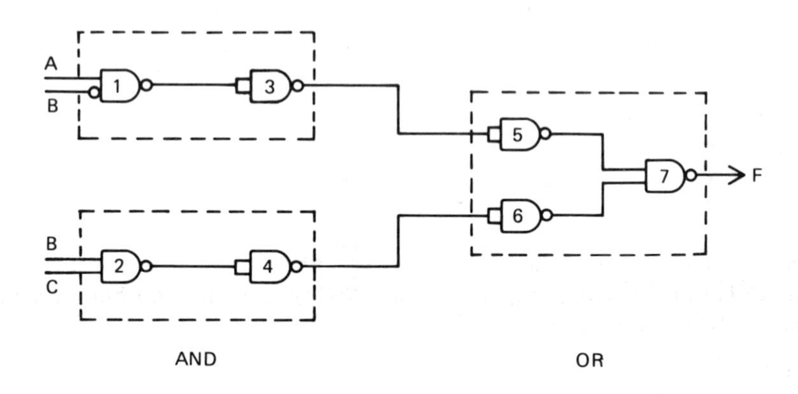

Gates 3 and 5 cancel out, as do 4 and 6.

If every propagation path through the NAND circuit is examined and double inversions removed, the circuit becomes

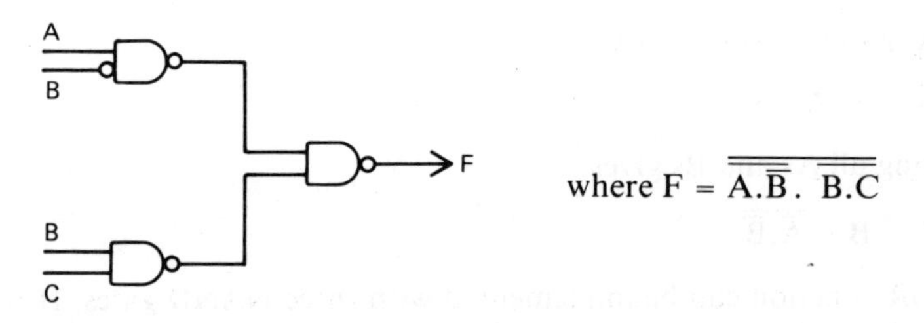

where $F = \overline{\overline{A.\bar{B}}\,.\,\overline{B.C}}$

By comparing the final NAND circuit with the original AND/OR version, the reader can see that the structure and interconnections are identical. The only difference between them is that every gate in the AND/OR version has been replaced with a NAND gate.

In a 2-level 1st canonical form circuit, each input variable or its inverse first passes through an AND gate (the 1st level of logic) whose output is then input to an OR gate (the 2nd level).

There is however an exception to this behaviour. If the 1st canonical form is not completely 2-level and any input does not pass through an AND gate, it must be inverted before the AND/OR gates are replaced with NANDS.

Take, for example, the function

$$F = A + BC$$

which is a very simple 1st canonical form, where variable A goes directly to the output OR gate. Its circuit is

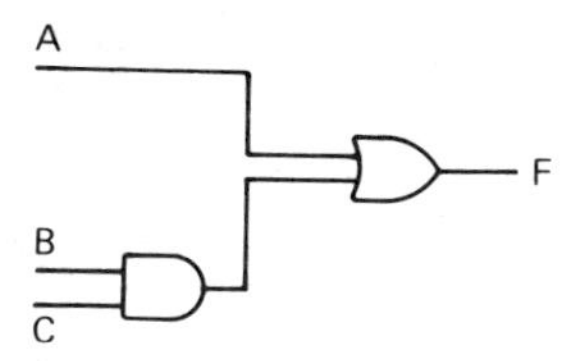

Direct replacement of the AND/OR gates with the NAND equivalent circuits gives

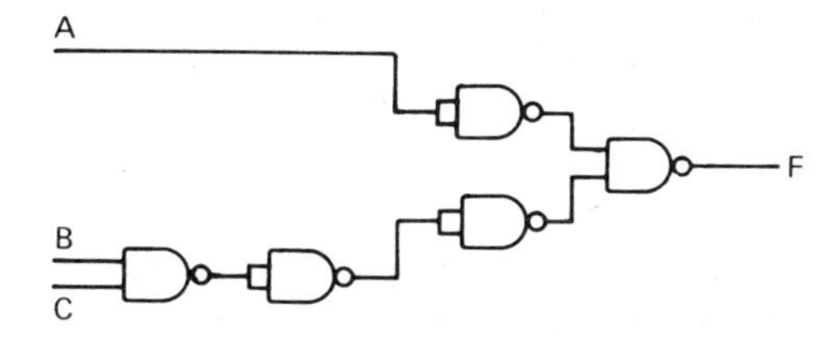

which simplifies to

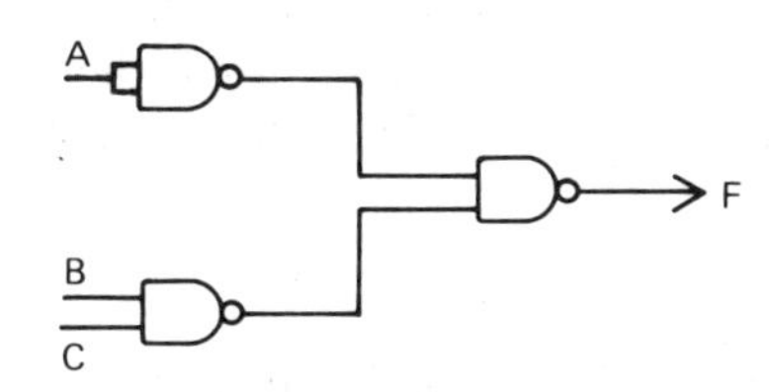

where $F = \overline{\overline{A}.\ \overline{B.C}}$

The A input must be inverted (with a NAND gate) before the AND/OR gates are replaced with NANDs, thereby satisfying De Morgan's theorem, which, when applied to F = A + BC gives

$$F = \overline{\overline{A}.\ \overline{B.C}}$$

The rule to obtain the NAND form is as follows. The NAND form of a combinational logic system can be obtained from the minimal 1st canonical form by replacing every gate with a NAND gate provided the original circuit is 2-level (AND/OR) throughout. If any variable or its inverse only passes through the output OR gate, an additional input inverter must be introduced into that line before all the gates are replaced with NANDs.

NOR Logic

NOR logic is the dual of NAND logic. AND, OR, and NOT functions can be performed with NOR gates only. The equivalent circuits are

NOT A F ≡ A F

OR A B F ≡ A B F

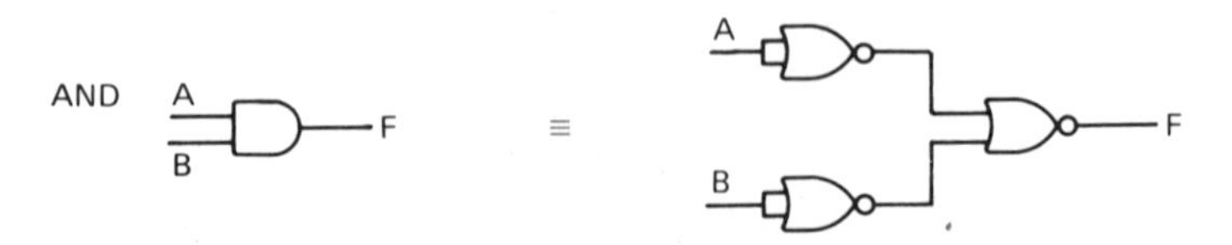

Remember NAND form is derived from 1st canonical form. NOR form is derived from 2nd canonical form.

The NOR form can be obtained from the minimal 2nd canonical form by replacing the OR/AND gates with NORs. If however a variable or its inverse only passes through the output AND gate, an additional inverter must be introduced on that input line before replacing the gates with NORs.

'Don't Care' Conditions

In some logic systems certain combinations of binary inputs may never occur. If they can be identified their corresponding outputs can be regarded as 'don't cares'. The designer is free to set 'don't care' outputs to either 0 or 1 and should choose the value which gives the simpler logic system.

Assume BCD is 8421 weighted. See Chapter 1.

For example, in BCD coding, where decimal digits are represented in pure binary, the numbers 1010_2 to 1111_2 are never used; therefore the outputs of any logic system being driven by BCD encoded data would be 'don't cares' for these inputs.

Worked Example 3.6 Obtain the minimal NAND and NOR logic functions to drive the top horizontal bar of a seven segment display from a BCD input.

[A 7-segment display comprises 7 light emitting bars a – g Each bar lights up when a logical 1 is applied to it. The illuminated bars form the shape of the decimal equivalent of the BCD input.]

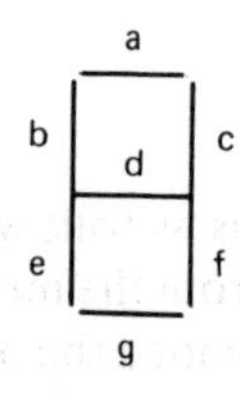

The logic function F_a, to drive bar a, is dependent on four binary inputs ABCD. Bar a must be illuminated and therefore $F_a = 1$ when the input has values 0, 2, 3, 5, 7, 8, and 9. Input values 10 to 15 give rise to output 'don't cares' as, in practice, these inputs will never occur.

The K-maps and truth table for F_a are

d stands for 'don't care'.

F_a CD \ AB	00	01	11	10
00	1_0	0_4	d_{12}	1_8
01	0_1	1_5	d_{13}	1_9
11	1_3	1_7	d_{15}	d_{11}
10	1_2	0_6	d_{14}	d_{10}

A	B	C	D	Fa
0	0	0	0	1
0	0	0	1	0
0	0	1	0	1
0	0	1	1	1
0	1	0	0	0
0	1	0	1	1
0	1	1	0	0
0	1	1	1	1

A	B	C	D	Fa
1	0	0	0	1
1	0	0	1	1
1	0	1	0	d
1	0	1	1	d
1	1	0	0	d
1	1	0	1	d
1	1	1	0	d
1	1	1	1	d

The NAND form is obtained from the 1st canonical form. The K-map looping is

Loop 1s for the 1st canonical form.

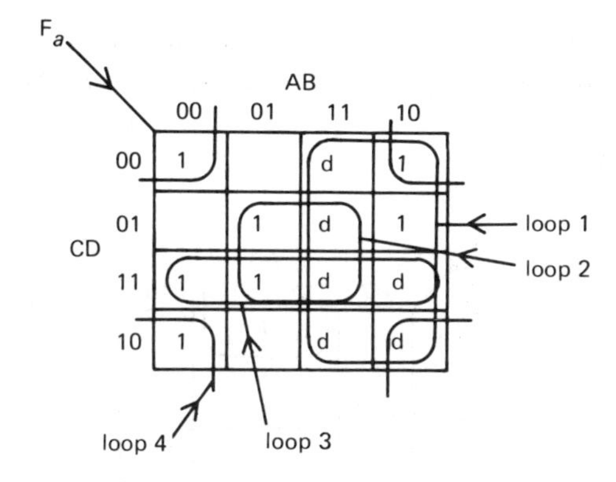

Loop 1 covers cells 8 and 9, but by setting all the 'don't cares' to 1, a loop of 8 cells can be drawn. Loop 2 covers cells 5 and 7, which together with 'don't cares' 13 and 14, give a loop of 4. Loop 3 covers cell 3, and loop 4 is a loop of 4 cells covering all the corners. Therefore

Note in particular loop 4. Side cells are adjacent as are the top and bottom cells.

$$F_a = \text{Loop 1 OR loop 2 OR loop 3 OR loop 4}$$
$$= A + BD + CD + \overline{B}\,\overline{D}$$

which can be converted to NAND form by inverting A and replacing each gate with NAND, giving

A only passes through the output OR gate and must therefore be inverted.

$$F_a = \overline{\overline{A}\,.\,\overline{(B.D)}\,.\,\overline{(C.D)}\,.\,\overline{(\overline{B}.\overline{D})}}$$

Thus the NAND circuit is

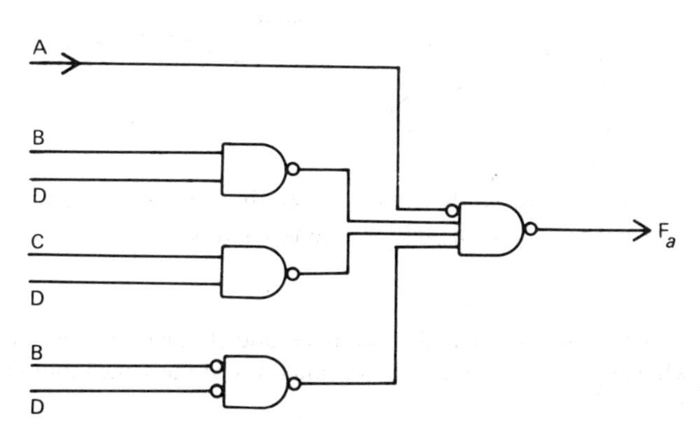

For the NOR version we require the 2nd canonical form. Redrawing the K-map for clarity and omitting the 1s gives

F_a

CD \ AB	00	01	11	10
00		0_4	d_{12}	
01	0_1		d_{13}	
11			d_{15}	d_{11}
10		0_6	d_{14}	d_{10}

There are 3 essential 0s to be covered. Cells 4 and 7 are adjacent and form a loop of 4 if 'don't cares' 12 and 14 are set to 0. Cell 1 has no adjacencies set to 0 and remains a loop of 1.

The looped K-map is

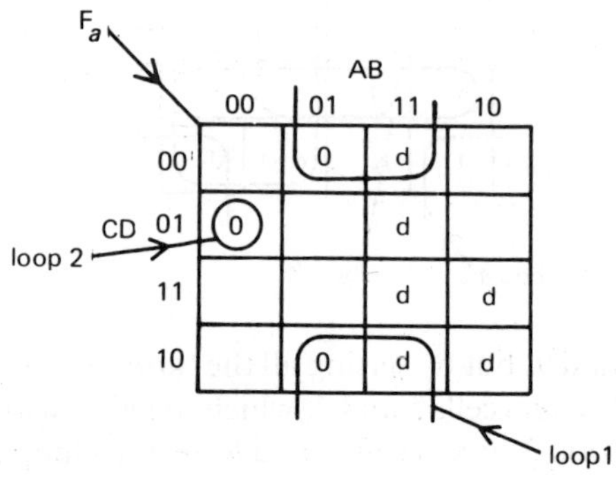

Hence $F_a = (\overline{B}+D).(A+B+C+\overline{D})$

This is a 2-level canonical form and every gate can be replaced with NOR giving

$$F_a = \overline{\overline{(\overline{B}+D)} + \overline{(A+B+C+\overline{D})}}$$

The NOR circuit is

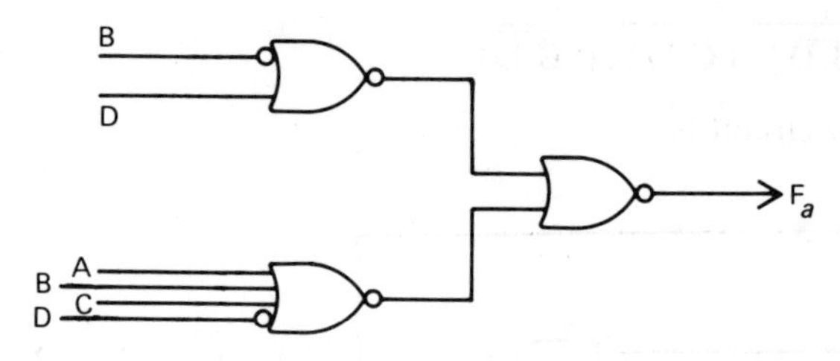

Both circuits also require two input inverters.

Comparing the NAND and NOR forms we see that the latter, requiring 3 gates, is simpler than the NAND version, which needs 4 gates.

Exercise 3.1 Complete the design of the 7-segment display driver and obtain the simplest version of functions F_b to F_g which drive bars b to g. All are functions of the same input variables A,B,C and D.

Hazards in Combinational Logic

Practical electronic logic circuits are not ideal devices. They require a finite time to operate and consequently introduce delays into the propagation of information. These delays are generally no more than a few microseconds and for very high speed logic, may be of the order of nanoseconds, but nevertheless they may invalidate the laws of Boolean algebra and cause errors or hazards in the logic state.

1 nanosecond equals 10^{-9} seconds

Consider the Boolean theorem

$$A.\overline{A} = 0 \tag{3.12}$$

The output should be permanently set to 0, regardless of the binary value of A. The logic circuit for this equation is

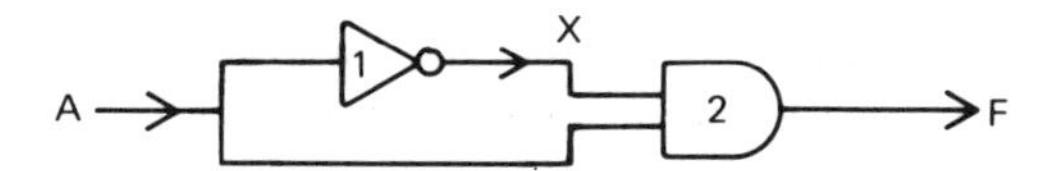

Let X be the output of inverter gate 1. If A is switched from 0 to 1, the direct line to the AND gate conveys this change immediately. The output (X) of the inverter, which also has A as input, will switch from 1 to 0, but it will take a finite time dt, to operate. The timing wave forms for the circuit are shown in Fig. 3.6. The output is not always at logical 0 as one would expect. At certain times the circuit contravenes Boolean algebra and $A.\overline{A} = 1$. This is known as a hazard, and results in a circuit malfunction.

Hazards are more likely to occur in multilevel logic circuits as the probability that different propagation paths through the system will have unequal delays, is greater. For this reason, canonical forms where there are at most, three levels of logic, including any input inverters, are preferable. Canonical circuits are, of course, not hazard-free, but the hazards are relatively simple both to detect and eliminate.

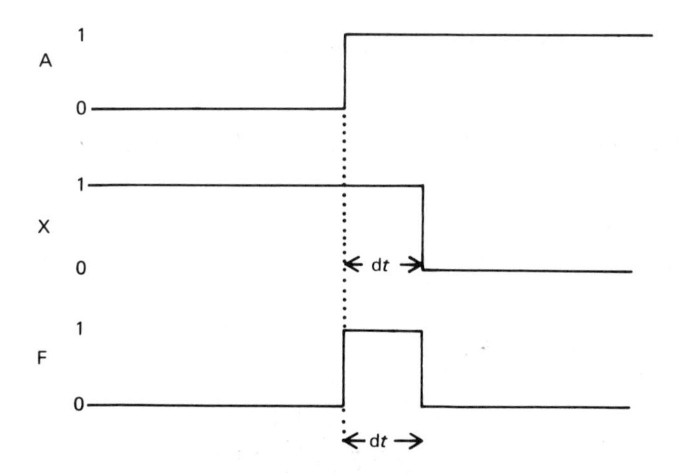

The AND gate is assumed to have zero delay.

Fig. 3.6 Timing diagram for circuit performing the function $F = A.\overline{A}$

Elimination of Hazards

There are three ways of eliminating hazards in combinational circuits. The first method is quite simply to wait. The correct output will always occur once the hazards have passed. If the maximum number of gates in any propagation path is n and the delay of each gate is dt, the output of a combinational logic circuit will always be valid at time T, after the input data has been applied, where

In practice T never need be greater than a few microseconds

$$T > n\,dt \tag{3.13}$$

This method is not suitable for combinational circuits whose outputs are used to drive a sequential system, because the incorrect outputs that occur when the hazards are present will become part of the input sequence to the sequential logic.

The second, rather ad hoc method, of eliminating hazards, is to try to balance out delays by using delay gate arrangements. An AND gate with common inputs will have no effect on the value of the data, but it will present a delay. Hence delays can be introduced deliberately into the propagation paths in a circuit to make them equal to the longest propagation delay.

Some delaying gate arrangements are given in Fig. 3.7.

A more rigorous technique for overcoming hazards involves the use of K-maps. A hazard will always occur when switching between adjacent cells on a K-map that are unlooped. The hazard can be removed by looping the adjacent cells even if it involves introducing an otherwise redundant term into the function.

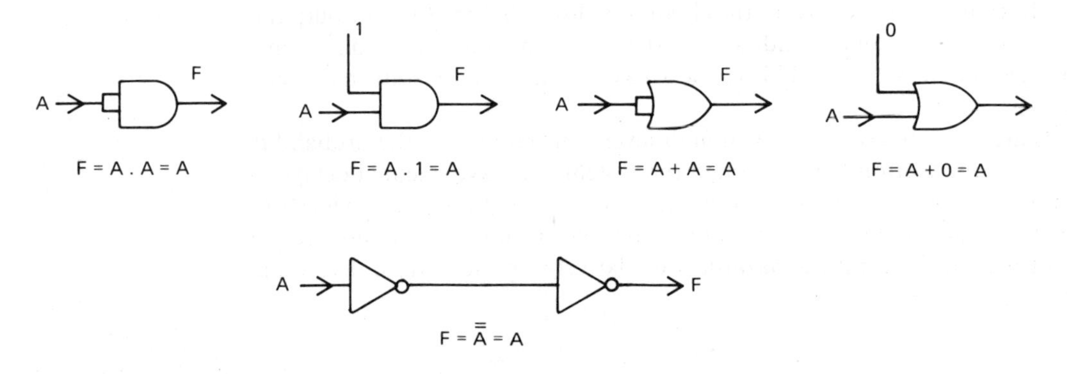

Fig. 3.7 Some delaying gate arrangements.

Worked Example 3.7 Identify and eliminate the hazard in the minimal 1st canonical form of the following function:

$$F = \Sigma\,(1,3,6,7)$$

The K-map is

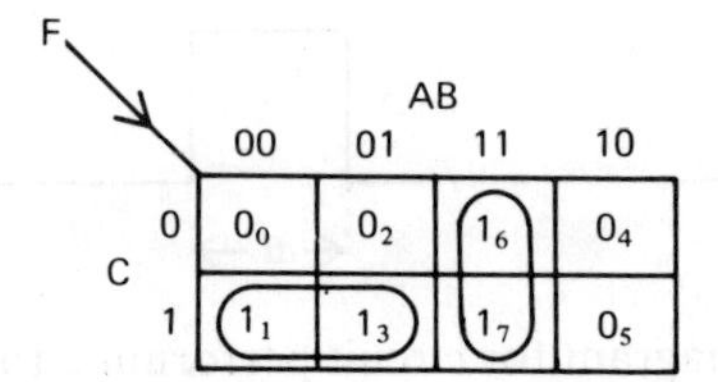

The minimal 1st canonical form is

$$F = AB + \overline{A}C$$

and the circuit

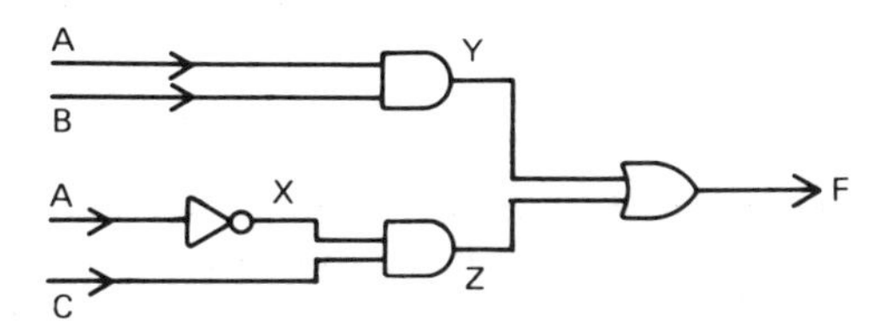

A hazard will occur when moving from cells 7 to 3. Inputs B and C are both at 1 and A switches from 1 to 0. There is an inverter in the propagation path A→X→Z→F and the resulting delay will cause F to switch momentarily to 0, whereas according to the algebra, it should remain at 1. The hazard may be removed by including the redundant loop (3,7) in the covering of the K-map. The circuit will then contain the term BC, which is independent of A and will hold the output F at 1, while A is switching.

The reader should verify this using a timing diagram.

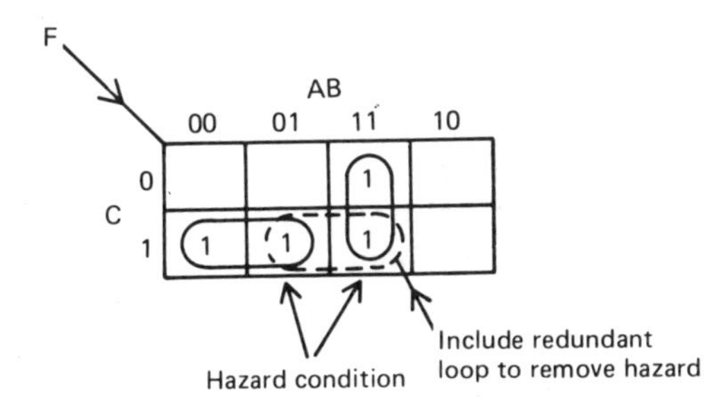

The hazard-free circuit is therefore

$$F = AB + \overline{A}C + BC$$

A K-map can be used to detect and eliminate all hazards arising from the switching of any one variable. Hazards can also arise if two or more variables switch instantaneously. However the likelihood of this happening with independent variables is negligible and maybe disregarded.

Tabular Method of Minimization

The K-map method of logic minimization is very useful when relatively small digital systems are designed. It is a very elegant 'pencil and paper' technique that avoids algebraic manipulations. If the number of variables is extended to 5 or 6, the map becomes three-dimensional and increasingly complex because of the potential number of overlapping loops. Beyond 6, the method breaks down. Furthermore, K-maps rely on the designer recognizing the largest possible loops and this human observation process is difficult to automate. Hence, the method cannot easily be programmed into a computer.

In three dimensions the loops become boxes containing 2^n cells.

The Quine – McCluskey algorithm involves the systematic and exhaustive reduction of a Boolean expression. It can handle any number of variables and is easily programmed into a digital computer, although minimization of simple functions can still be carried out by hand.

Use maxterms for minimal 2nd canonical form.

The starting point of the algorithm is the full list of minterms, if the minimal 1st canonical form is required.

Let F = Σ (0,4,10,11,12,13,14,15)

The first step in the algorithm is to group the minterms together according to the number of 1s contained in each input. F is a function of 4 variables, say W, X, Y and Z; therefore list 1 will have 5 groups — terms with no 1s, terms containing one 1, two 1s and so on.

m identifies the value of the minterm.

List 1

m	WXYZ	m	WXYZ	m	WXYZ	m	WXYZ	m	WXYZ
0	0000	4	0100	10	1010	11	1011	15	1111
				12	1100	13	1101		
						14	1110		
0 × 1s		1 × 1		2 × 1s		3 × 1s		4 × 1s	

This process is equivalent to applying the theorem $AB + A\bar{B} = A$ to every pair of minterms. A can be a single variable or a logic function.

List 2 is obtained by comparing all pairs of entries in adjacent groups in list 1. If a pair differs by 1 bit only, they are combined and the differing variable replaced with a d (for 'don't care'). List 2 is also grouped according to the number of 1s in the pairs, excluding the 'don't care' condition.

All pairs of minterms are equivalent to all possible loops of 2 on a K-map of the function.

List 2

m	WXYZ	m	WXYZ	m	WXYZ	m	WXYZ
0,4	0d00	4,12	d100	10,11	101d	11,15	1d11
				10,14	1d10	13,15	11d1
				12,13	110d	14,15	111d
				12,14	11d0		
0 × 1s		1 × 1		2 × 1s		3 × 1s	

An implicant is formed when minterms are combined together. If a minterm cannot be combined it is also an implicant.

Lists 1 and 2 are compared to see if every minterm in list 1 is carried through and appears somewhere in list 2. If a term does not carry through, it cannot be combined further. It is an implicant of the function. In this example every term in list 1 carries through to list 2.

List 3 is formed by combining entries in adjacent groups in list 2. Entries may be combined if they differ by 1 variable and have the same variable eliminated — the 'don't cares' must be common.

The order of the minterms is unimportant. Term 10,14,11,15 is identical to 10,11,14, 15 and is therefore omitted.

List 3

m	W X Y Z
10,11,14,15	1 d 1 d
12,13,14,15	1 1 d d
2 × 1s	

The implicants in List 2 which do not carry through to list 3 are (0,4) and (4,12).

Equivalent to loops of 4 on a K-map.

Subsequent lists are calculated until no further combination of terms is possible. In this example, list 3 is the final list and terms (10,11,14,15) and (12,13,14,15) are also implicants of the function.

The set of implicants covers the logic function, but like the loops on a K-map, some terms may be redundant.

Prime Implicant Table

In order to ascertain the essential or prime implicants, the redundant terms must be identified. This can be done on a prime implicant table, which has columns labelled with the original minterms of the function and rows with the implicants. A flag is inserted at the intersection of rows and columns if the minterm (column label) is contained in the implicant (row label).

We have already identified the implicants (0,4), (4,12), (10,11,14,15,) and (12,13,14,15) in the function F = Σ (0,4,10,11,12,13,14,15). Its prime implicant table is given in Table 3.3.

Table 3.3 Prime Implicant Table for

F = Σ (0, 4, 10, 11, 12, 13, 14, 15)

		Original minterms							
Implicants		0	4	10	11	12	13	14	15
(0, 4)	(0d00)	*	*						
(4, 12)	(d100)		*			*			
(10, 11, 14, 15)	(1d1d)			*	*			*	*
(12, 13, 14, 15)	(11dd)					*	*	*	*

By scanning the table, the columns containing only one star, identify the essential prime implicants. (0,4) is needed to cover minterm 0 (it also covers 4); (10,11,14,15) to cover 10 and 11; (12,13,14,15) to cover 13. These three prime implicants cover all the original minterms and (4,12) is therefore redundant.

The logic function can be obtained from the binary values of the essential prime implicants giving

$$F = \overline{W}\,\overline{Y}\,\overline{Z} + WY + WX$$

From list 2,(0,4) represents WXYZ = 0d00 and can therefore be expressed as $\overline{W}\,\overline{Y}\,\overline{Z}$.

The Quine - McCluskey method, although longer than the K-map minimization, can easily be programmed into a computer and may be extended to any number of variables. (The reader should minimize the previous function on a K-map and show that it gives the same result).

Exercise 3.2

Using the Quine - McCluskey algorithm minimize the following functions
(i) F = Σ(3,4,5,7,9,13,14,15)
(ii) F = Σ(1,2,3,6,8,9,10,11,18,21,22,24,25,26,27)
(If you have a computer, you may wish to write a program to solve ii)

Cellular Logic

In some logic systems, particularly where there are a large number of input variables, the canonical forms become increasingly complex and unwieldy. Furthermore, any minor change in the specification of the system often leads to a major redesign of the circuit.

An alternative design approach is to decompose the logic function into a series of identical operations that can be carried out on part of the input data. If a logic cell can be identified and designed, then the complete system may be built up by connecting a number of cells together.

A general cellular circuit is shown in Fig. 3.8. Each cell receives external inputs $\mathbf{I}_i$ and generates internal outputs $\mathbf{Q}_i$ which are input to an adjacent cell. There may be external outputs at each cell or the output may be derived from the boundary cells, depending on the problem. The direction of flow of information between cells is deliberately not shown in Fig. 3.8 as it is problem dependent. Data flow may be from least to most significant cell or vice versa. In certain other problems it could be either way. The complexity of each cell depends on the number of external inputs it receives, but a trade-off can be made as the larger the number of inputs to a cell, the fewer the cells required in the system.

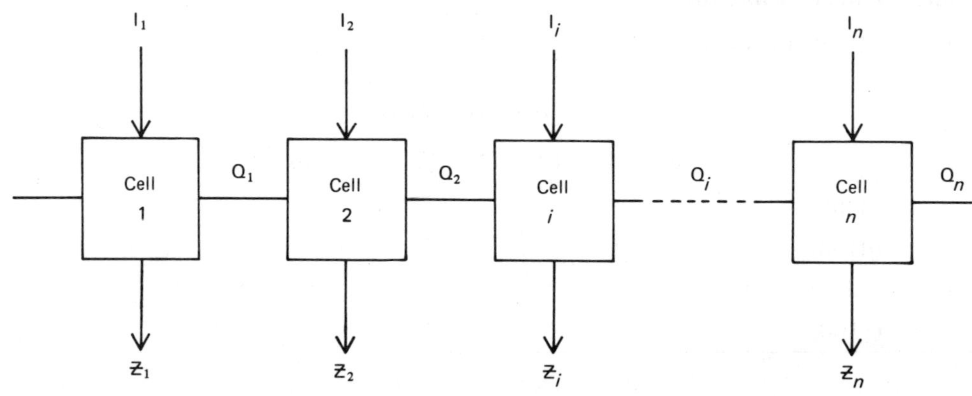

Fig. 3.8 A cellular circuit.

The design of an adder circuit illustrates the essential steps necessary to produce a cellular logic system. Consider a 2-bit adder with inputs $\mathbf{A} = A_1 A_0$ and $\mathbf{B} = B_1 B_0$ and outputs $\mathbf{S} = S_2 S_1 S_0$, where

Beware. In this Equation, + means add.

$$\mathbf{S} = \mathbf{A} + \mathbf{B}$$

Firstly, for the purpose of comparison, let us obtain the canonical form of the adder. The output consists of three functions S_2, S_1 and S_0 whose truth tables can be evaluated by considering the sum of all possible values of A and B. The K-maps of these truth tables are:

There are 16 different input conditions $A_1\ A_0\ B_1\ B_0$.

Assume the outputs are functions of all the input variables.

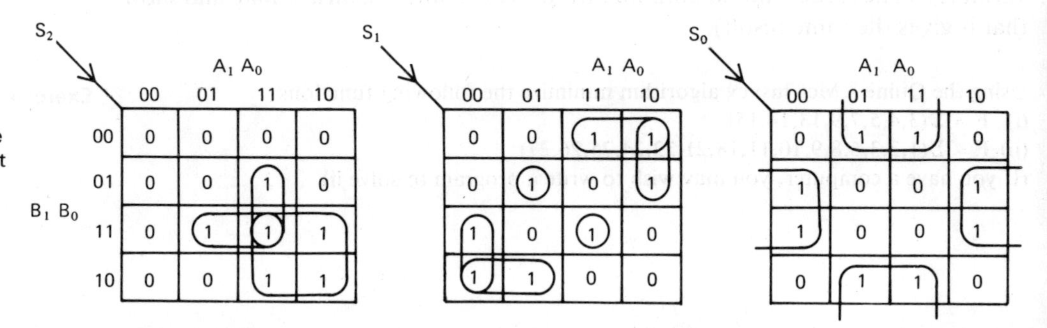

iving $S_0 = A_0 \overline{B}_0 + \overline{A}_0 B_0 = A_0 \oplus B_0$

$S_1 = A_1 \overline{B}_1 \overline{B}_0 + A_1 \overline{A}_0 \overline{B}_1 + \overline{A}_1 \overline{A}_0 B_1 + \overline{A}_1 B_1 B_0 + \overline{A}_1 A_0 \overline{B}_1 B_0 + A_1 A_0 B_1 B_0$

ad $S_2 = A_1 B_1 + A_0 B_1 B_0 + A_1 A_0 B_0$

he circuits operate in parallel on common inputs, so the maximum propagation elay will, at most, be three gate delays.

Before designing a cellular circuit for the adder, we must first devise a suitable ructure. Let us consider how we, as humans, perform addition. We start with the ast significant bits of the input words. Taking one bit from each word and any rry from the previous stage, we add them together to produce an output sum bit, d a carry which is input to the next stage of the addition. This procedure is peated for every binary place in the addition. The adder cell is therefore:

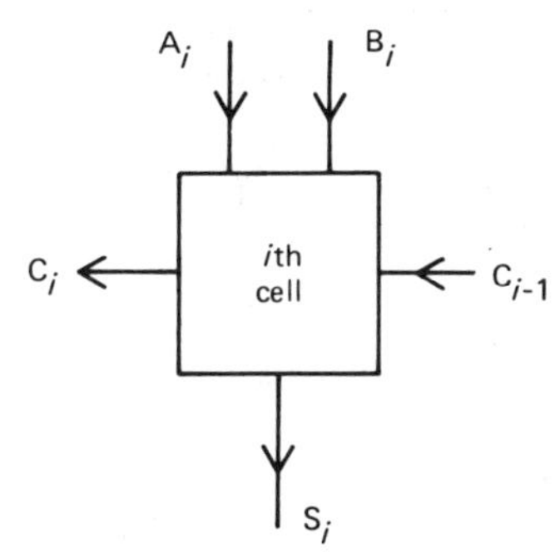

here A_i and B_i are the external inputs, being the ith bits of the input words **A** and respectively. C_{i-1} is the carry-in from the previous cell, C_i is the carry-out to the + 1)th cell and S_i is an external output forming the ith bit of the sum.

Functions C_i and S_i can now be designed. They are combinational logic functions three input variables A_i, B_i and C_{i-1} and represent the sum and carry of the binary dition of three bits.

The truth table is

A_i	B_i	C_{i-1}	C_i	S_i
0	0	0	0	0
0	0	1	0	1
0	1	0	0	1
0	1	1	1	0
	0	0	0	1
	0	1	1	0
	1	0	1	0
	1	1	1	1

ving K-maps

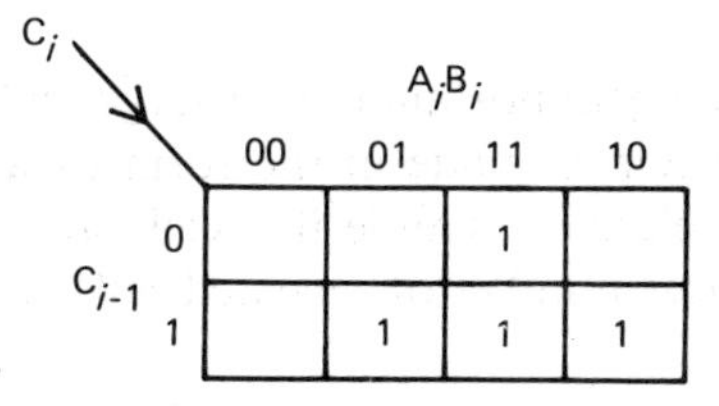

S_i

C_{i-1} \ A_iB_i	00	01	11	10
0		1		1
1	1		1	

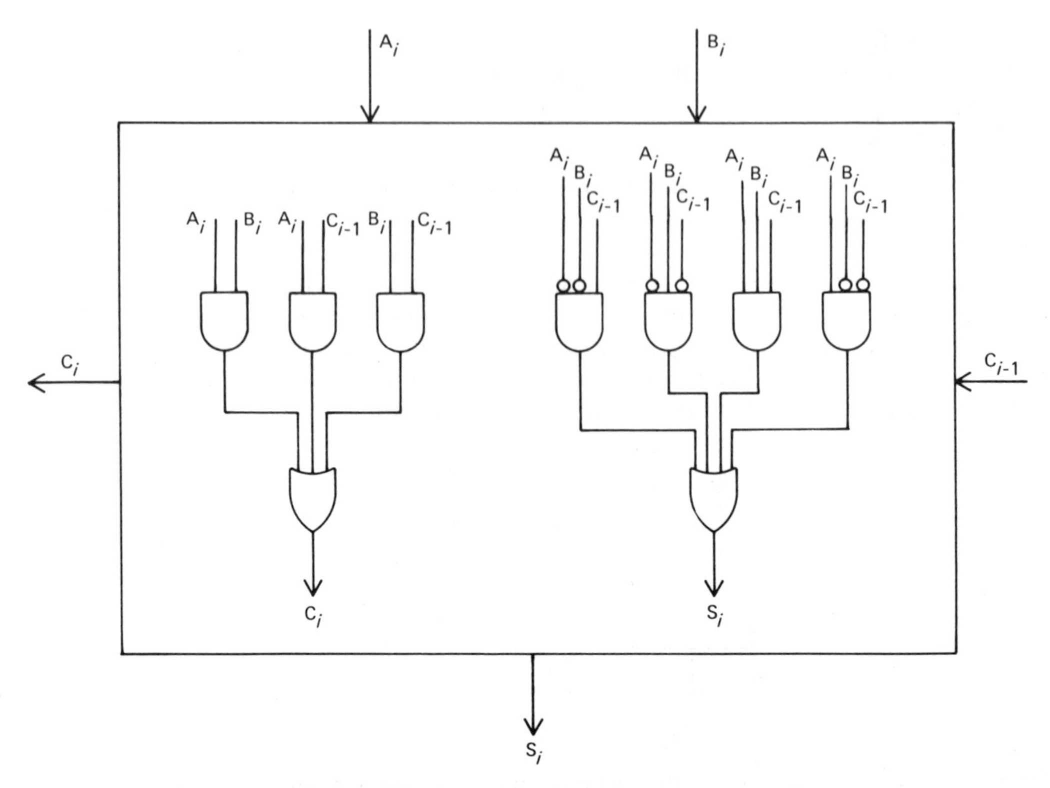

Fig. 3.9 A full adder cell.

S_i does not simplify as a canonical form. If EX.OR gates are available it can be simplified by algebra to

$$S_i = A_i \oplus (B_i \oplus C_{i-1}).$$

If the cell is designed with only one external input it is known as a half adder. Twice as many cells would be required compared with full adder circuits.

and functions

$$C_i = A_i B_i + A_i C_{i-1} + B_i C_{i-1}$$

and $$S_i = \overline{A}_i \overline{B}_i C_{i-1} + \overline{A}_i B_i \overline{C}_{i-1} + A_i B_i C_{i-1} + A_i \overline{B}_i \overline{C}_{i-1}$$

The circuit diagram for the cell is shown in Fig. 3.9. This is often called a full adder and two cells would be required for the 2-bit adder. Some boundary conditions may have to be applied to the end cells. The carry input at cell 0 must be 0 as there cannot be a carry before the least significant bits have been processed. The carry-out from the most significant stage becomes the final sum bit, hence, $S_2 = C_1$. The block diagram of a 2-bit cellular adder is given in Fig. 3.10.

The main advantage of the cellular structure over the canonical form is that the former can be expanded by simply adding on extra cells. A 3-bit adder would need one extra cell whereas a 16-bit adder would require a further thirteen. The cells are all identical — their functions do not change with the increasing resolution of the adder.

Compared, with the canonical approach, a 3-bit adder would require a complete redesign of the function S_2 and a new function (S_3) of six input variables. In the case of a 16-bit canonical form adder, the most significant sum bit would be an unwieldy function of 32 input variables.

Iterative is an alternative name to cellular.

The cellular or iterative logic structure is extremely important in very large scale integration manufacture. If a complex logic system is made in the form of an integrated circuit, the design process becomes relatively simple if a cell can be devised. The complex system then comprises a large number of repeated cells and

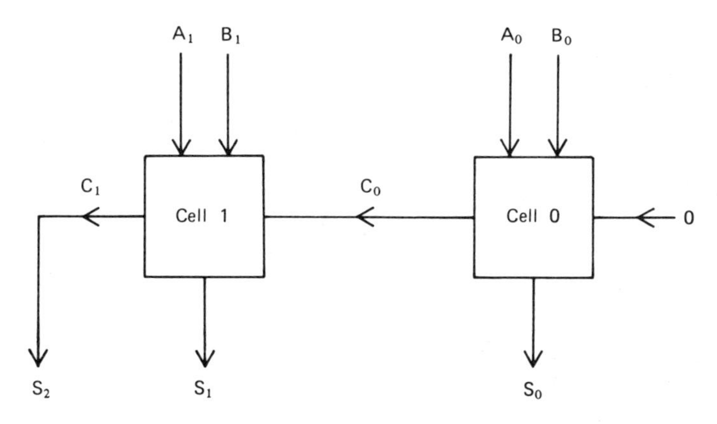

Fig. 3.10 A 2-bit adder circuit.

this may be carried out photographically on the integrated circuit mask or by using a computer-aided drawing package.

The main disadvantage of the cellular form of logic is its operating time, which is directly proportional to the number of cells in the system. In the canonical form of the adder the maximum delay was 3d*t* and this is independent of the resolution of the adder, as all functions effectively operate in parallel from common inputs. In the cellular circuit each cell has a delay of 3d*t*. Taking the worst case delay in the adder, where each cell generates a carry, a period of 3*n* d*t* (where *n* is the number of cells in the system) must elapse before the output is valid, otherwise hazards may be present on the output.

d*t* is the delay due to one gate.

Cells operate in series — delays are cumulative.

Worked Example 3.8

Design a cellular circuit to generate the odd parity for 4-bit words and compare it with its equivalent canonical form.

Odd parity — an additional bit in a binary word which makes the total number of bits set to 1, an odd number.

System structure

1. *Data flow.* As parity can be calculated either by starting with the least or most significant bit, the direction of data flow does not matter in this problem.
2. *Cell specification.* Let each cell process 1 bit of the input word and calculate P_i — the 'parity so far'. The final output P is the 'parity so far' after all the inputs have been processed. No external output is required at each cell.

The block diagram for the cell is

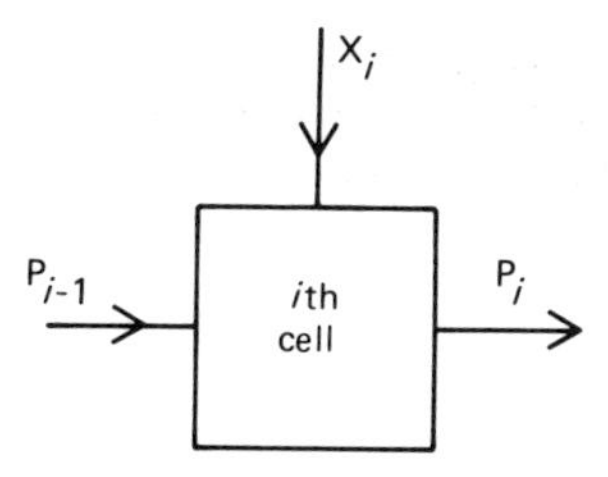

If the parity (P_{i-1}) for inputs X_0 to X_{i-1} is 0, the number of bits set to 1, between bits X_0 to X_{i-1}, must be odd. If the current input X_i is 0, the new 'parity so far' (P_i) needs to be 0. The parity for all other cell input conditions can also be calculated and the cell truth table obtained, which is

P_{i-1}	X_i	P_i
0	0	0
0	1	1
1	0	1
1	1	0

P is therefore the EXCLUSIVE-OR function.

$$P_i = P_{i-1} \oplus X_i$$

so the cell is simply an EX.OR gate.

The complete parity generator is therefore

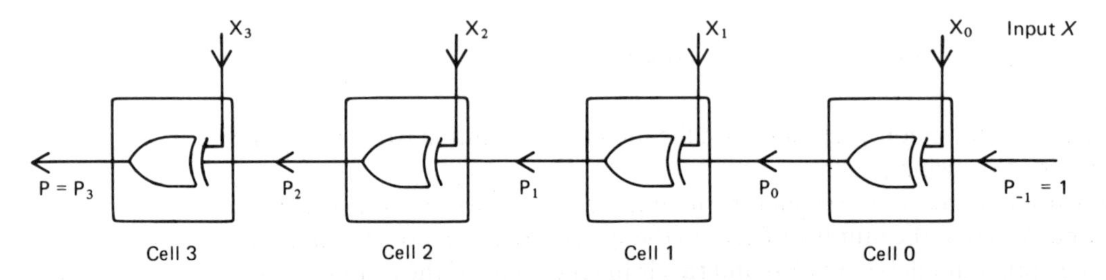

The canonical form of the odd parity generator can be obtained directly from the K-map specification.

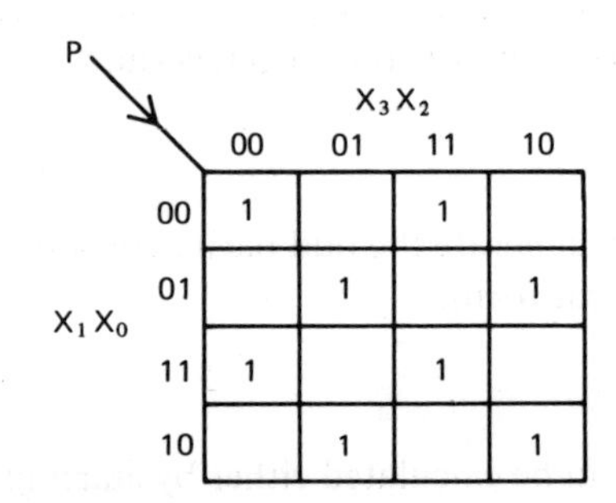

X_1X_0 \ X_3X_2	00	01	11	10
00	1		1	
01		1		1
11	1		1	
10		1		1

It does not minimize and is therefore

$$P = \overline{X}_3\overline{X}_2\overline{X}_1\overline{X}_0 + \overline{X}_3\overline{X}_2X_1X_0 + \overline{X}_3X_2\overline{X}_1X_0 + \overline{X}_3X_2X_1\overline{X}_0 + X_3\overline{X}_2\overline{X}_1X_0 + X_3\overline{X}_2X_1\overline{X}_0 + X_3X_2\overline{X}_1\overline{X}_0 + X_3X_2X_1X_0$$

Exercise 3.3 Design a cellular circuit to indicate single errors in a '2 in 5' encoder. (Hint: 2 bits of information must be passed between cells. Let Q = 00 be '0 bits set so far', 01 be '1 bit set', 10 be '2 bits set' and 11 be 'more than 2 bits set'. An error is present if output from the final cell is NOT 10)

Summary

This chapter has been concerned with the fundamental principles of combinational logic design. We have seen that a logic function is dependent on the definition of the logic levels on the inputs and outputs. If the coding is inverted, the dual function is obtained.

Two standard logic forms have been defined. The first canonical form has an AND level of logic immediately after the inputs followed by an OR gate, whereas the 2nd form is OR followed by AND. Both the 1st and 2nd canonical forms can be obtained from a truth table and they both perform the same logical operation.

Canonical forms can be minimized. The K-map method is suitable for optimizing small logic systems up to a maximum of 6 variables. A computerized version of the Quine–McCluskey algorithm may be used for complex systems.

Any logic system can be constructed entirely of NAND gates or NOR gates. A 1st canonical form function, provided it is two-level throughout, may be converted to NAND form by replacing every gate with a NAND. The 2nd form can be similarly converted to NOR only.

In some logic systems, 'don't care' conditions can be identified on the outputs. The designer is free to set a 'don't care' output to either 0 or 1, and this may influence his minimization.

Electronic logic circuits are imperfect devices. They do not operate instantaneously. The delays they introduce, momentarily, may cause the laws of Boolean algebra to be violated and the circuit to malfunction. Methods of removing these hazards have been examined in this chapter.

Finally, a cellular structure has been proposed that has both advantages and disadvantages compared with the canonical form. A cellular structure is however ideally suited to integrated circuit manufacturing techniques.

Problems

3.1 The functions of circuits within a logic system with positive logic coding are:

$$F_1 = \overline{A}\,\overline{B}\,\overline{C} + \overline{A}BC + A\overline{B}\,\overline{C}$$

& $$F_2 = A\overline{B}C + AB\overline{C} + ABC$$

What functions do the circuits perform if negative logic coding is used?

3.2 (a) Obtain the minimal 1st canonical forms for the following equations:

$$F_1 = \overline{A}\,\overline{B}\,\overline{C} + \overline{A}\,\overline{B}C + \overline{A}B\overline{C} + A\overline{B}C$$
$$F_2\,(ABCD) = \Sigma(3,4,5,7,9,13,14,15)$$
$$F_3\,(ABC) = \Pi(2,3,4,5,6,7)$$

(b) Obtain the minimal 2nd canonical form for the above equations.

3.3 Obtain the minimal NAND forms of the following functions:

$$F_1 = \Sigma(3,4,5,6,7,8,10,12,13)$$
$$F_2 = \Sigma(0,1,2,3,4,5,8,9,10,11)$$

3.4 Obtain the minimal NOR forms of the following functions:

$$F_1 = \Sigma(0,2,3,4,6)$$
$$F_2 = \Pi(0,1,2,8,9,12,13,14,15)$$

3.5 Obtain the minimal 1st and 2nd canonical forms of the following function:

$$F = \Sigma(4,5,6,8,10,14)$$

& 'don't cares' $= \Sigma(1,3,7,12,15)$

3.6 Design a digital system using only the minimum number of NAND gates, that will convert 8,4,2,1 (ABCD) into 8,4, −2, −1 (WXYZ) binary coded decimal.

3.7 Design a digital system using AND/OR/NOT gates that will convert 4-bit binary (ABCD) into Gray code (WXYZ). What would be the minimum number of gates required if EXCLUSIVE-OR gates were made available?

3.8 Identify and eliminate the hazards in the following circuit:

$$F = \Sigma(1,2,3,5,12,13,14,15)$$

3.9 Minimize the following function, by means of the Quine – McCluskey algorithm:

$$F = \Sigma(1,4,6,7,12,13,14,15,17,28,29,30,31)$$

3.10 Design a cellular logic system that will accept 8-bit words *A* and *B* and indicate whether *A* is greater than, equal to, or less than *B*.

Sequential Logic Fundamentals 4

Objectives

- ☐ To investigate the structure and behaviour of a simple sequential circuit.
- ☐ To examine the set–reset flip-flop, its properties and limitations.
- ☐ To examine the JK flip-flop and its derivatives — the master–slave device, the synchronous and asynchronous triggers and the delay flip-flop.
- ☐ To investigate counting circuits based on asynchronous trigger flip-flops.
- ☐ To explain the structure of a shift register.
- ☐ To determine the effects of feedback on a shift register.

Feedback is an essential feature of many natural and artificial systems

Having examined combinational logic in detail in Chapter 3, we now investigate the effects of applying some feedback to a logic circuit. The feedback creates a sequential system whose output is dependent not only on the present input, but also on all the previous inputs over a given sequence. Another property of a sequential circuit is its ability to give a different output response when the same input is re-applied to the circuit. This behaviour cannot happen in a combinational circuit. In this chapter a simple sequential circuit with a single feedback connection will be used as a prototype from which a range of flip-flops will be developed. The flip-flops can be regarded as the building block of a discrete sequential logic system.

A Sequential Logic Circuit

A sequential logic circuit, in its simplest form, is a combinational logic system with some feedback connections from the output, providing one or more additional internal inputs. Data is applied to the external inputs, but the user has no direct control over the internal inputs. They can only be changed by the logic circuit itself responding to all its inputs, both external and internal. The value of the internal inputs defines the state of the system.

A sequential logic circuit comprising a combinational system with some feedback is shown in Fig. 4.1. It has a single output Q which is also fed back to form internal input Q′ (the state of the system). The external inputs are S and R and the system can be described by the Boolean equation

$$Q = \overline{R + \overline{(S + Q')}} \qquad (4.1)$$

where Q, on being fed back, becomes Q′.

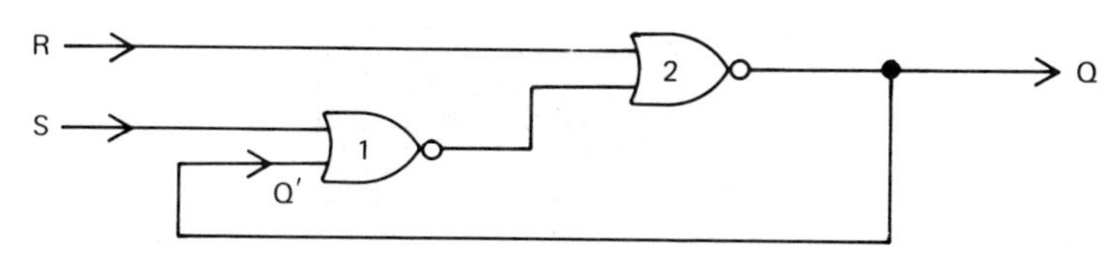

Fig. 4.1 A sequential circuit formed by applying feedback to a combinational circuit.

Table 4.1 Truth Table for Sequential Circuit
$Q = R + \overline{(S + Q')}$ with Feedback Loop Open Circuit

S	R	Q′	Q	Stability	
0	0	0	0	Stable	①
0	0	1	1	Stable	②
0	1	0	0	Stable	③
0	1	1	0	Unstable	
1	0	0	1	Unstable	
1	0	1	1	Stable	④
1	1	0	0	Stable	⑤
1	1	1	0	Unstable	

'Inputs' are the external inputs.

The output is determined by both the input and state values, so it is possible to obtain a different response if the same inputs are re-applied. The present output at time *t* is a function of the present inputs and state, but the latter is a function of inputs and state at time *t*-1. The state at *t*-1 depends on inputs and state at *t*-2 and so on. The present output is therefore dependent on the present input and all previous inputs which have been applied to the system. This property characterizes a sequential system.

Sometimes called a Huffman switch, after D.A. Huffman who first suggested the idea.

A feedback logic sequential circuit may be analysed by imagining that there is a switch in the feedback loop that isolates Q from Q′. If the switch is open, the circuit becomes a combinational logic system and its truth table can be calculated. The truth table for the feedback circuit in Fig. 4.1 is given in Table 4.1.

When the Huffman switch is closed, Q′ is forced to take the value of Q. If Q has the same value as Q′, nothing changes and the circuit is said to be STABLE. If however, Q changes Q′, a different set of inputs is now applied to the logic because the state has changed even though the external inputs R and S are held constant. The circuit is therefore UNSTABLE and a new output will be generated. The circuit will continue to switch until a stable state is entered.

If no stable state exists, the circuit will oscillate between unstable states.

The stability of a circuit can be determined from its truth table:

If $Q = Q'$ then the circuit is stable
If $Q \neq Q'$ the circuit is unstable

The stability conditions are given in Table 4.1 alongside the output data and can be summarized on a K-map where each stable cell is labelled with a circled number (e.g., ①), giving

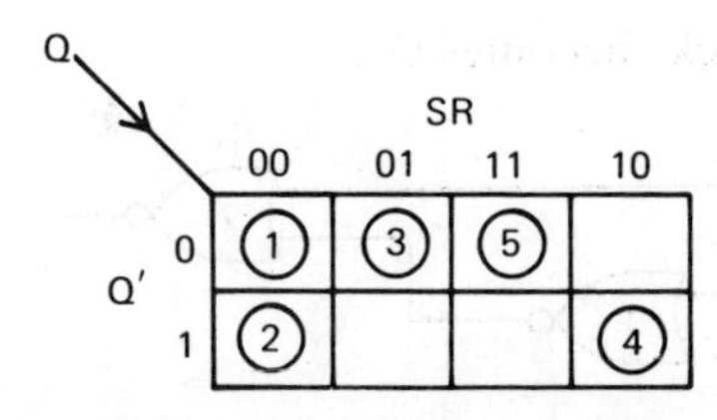

The unstable states may now be examined. Consider SR = 01 and Q′ = 1. The output Q becomes 0 and will, when fed back, switch Q′ to 0, thereby creating the

input conditions for stable state ③. Therefore the circuit state with SR Q' equal to 011 is unstable state 3, which will switch to stable ③. Unstable states 4 and 5 can also be identified. The complete stability map is called a flow table, and is given in Table 4.2.

Table 4.2 Truth Table for the Sequential Circuit $Q = \overline{R + \overline{(S + Q')}}$

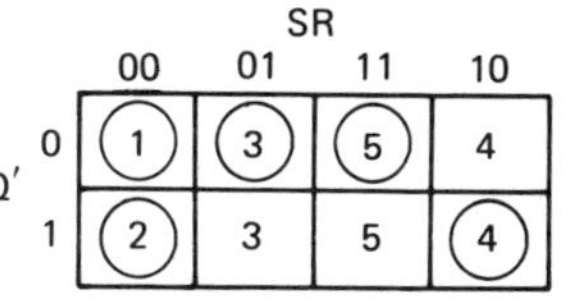

Q' \ SR	00	01	11	10
0	①	③	⑤	4
1	②	3	5	④

Stable states are circled and their corresponding unstable states have the same reference number, but are uncircled.

Flow tables are only used to summarize the stability of a circuit. They are not used for minimization; therefore the axes do not necessarily have to be Gray coded, and are not restricted to two variables. The columns are however always labelled with the external inputs and the rows with the internal states. Any horizontal motion within the flow table can be brought about by changing the external inputs, whereas the vertical motion is controlled by the stability of the circuit.

Gray coding is convenient if information is being transferred from a K-map to a flow table.

A flow table can be used to investigate the behaviour of a circuit. In state ① (Table 4.2) the circuit is stable and S and R are both 0. If S is set to 1, unstable state 4 will be entered. The circuit will then switch itself into stable ④ and both the internal state Q' and the output will become 1. If SR is returned to 00, state ② is entered and the output remains at 1, whereas previously with SR at 00 the output was 0. If R is now set to 1, state ③ is entered and the output becomes 0. The circuit is a 1-bit memory. It will hold the value 1 on its output if S is momentarily set to 1. The memory can be cleared to 0 if the R input becomes 1. This sequential circuit is known as a Set–Reset Flip–Flop (SRFF) where S is the Set input and R is the Reset input.

Flip-flops are sometimes called bistables.

Worked Example 4.1

Investigate the stability of the following system and obtain its flow table

$$F_1 = \overline{A}\,\overline{F_1'} + \overline{B}F_1' + \overline{A}\,\overline{F_2'} + AF_1'F_2'$$

and $$F_2 = \overline{A}\,\overline{B} + \overline{A}F_2' + \overline{B}\,\overline{F_1'}F_2' + BF_1'F_2'$$

where $F_1 \rightarrow F_1'$ and $F_2 \rightarrow F_2'$

$F_1 \rightarrow F_1'$ reads F_1 is fedback to input F_1'.

First obtain the truth tables of F_1 and F_2. Both are functions of 4 variables, A, B, F_1' and F_2'

A	B	F_1'	F_2'	F_1	F_2	Stable state
0	0	0	0	1	1	
0	0	0	1	1	1	
0	0	1	0	1	1	
0	0	1	1	1	1	①
0	1	0	0	1	0	
0	1	0	1	1	1	

(cont'd on page 62)

A	B	F_1'	F_2'	F_1	F_2	Stable state
0	1	1	0	1	0	②
0	1	1	1	0	1	
1	0	0	0	0	0	③
1	0	0	1	0	1	④
1	0	1	0	1	0	⑤
1	0	1	1	1	0	
1	1	0	0	0	0	⑥
1	1	0	1	0	0	
1	1	1	0	0	0	
1	1	1	1	1	1	⑦

By comparing $F_1' F_2'$ with $F_1 F_2$, the stable states can be identified. This circuit is only stable if both F_1' and F_2' are identical to F_1 and F_2. Seven stable states can be identified and entered onto a flow table.

Now considering the unstable states we have, for inputs AB = 00 and $F_1' F_2' = 00$, outputs $F_1 F_2 = 11$. After feedback $F_1'F_2'$ becomes 11 and this is the input condition for stable state ①. Hence $ABF_1' F_2' = 0000$ is unstable state 1.

The complete flow table is

Internal states $F_1' F_2'$ \ External inputs AB	00	01	11	10
00	1	2	⑥	③
01	1	↑ 8a	6	④
11	①	8b ↓	⑦	5
10	1	②	6	⑤

Oscillations are indicated by arrows on the flow table.

It is left as an exercise for the reader to check for him/herself all the remaining unstable states. Note in particular the oscillation between unstable states 8a and 8b and that stable state ① has three unstable states associated with it.

Limitations of the Set–Reset Flip-Flop

Flow tables represent ideal circuit behaviour. They do not take into account any timing problems. Suppose the SR input data on an SRFF changes from 11 to 00 (See Table 4.2). One would expect the circuit to switch from ⑤ to ① and the output to remain at 0. In practice, S and R will not both change at the same instant of time. If S switches before R, the circuit will enter ③ and then ① and the output will remain at 0. If however, R becomes 0 before S, we leave ⑤ and enter unstable state 4. The circuit will automatically switch to stable ④ and then to ② when S becomes 0. The output is now 1. The behaviour of the circuit is therefore uncertain and depends on whether S or R switches first. It will be sensitive to very small time differences between the two inputs.

Typical value is 1 microsecond.

The problem would appear to be soluble by electronically synchronizing the inputs. Unfortunately this will reveal another timing problem. Figure 4.2 shows the data values on all gate inputs and outputs when an SRFF is in state ⑤. If S and R

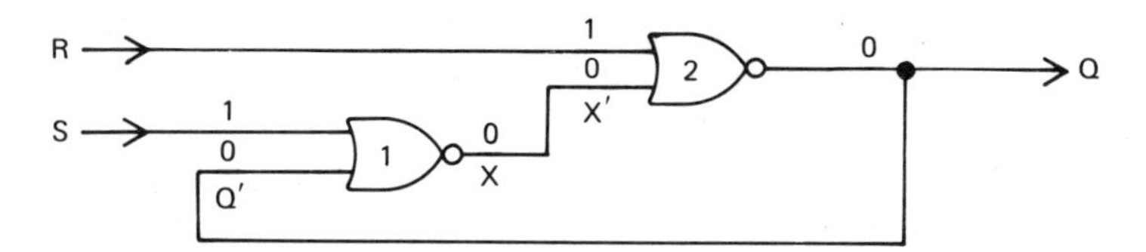

Fig. 4.2 Data values in a SRFF when SR = 11.

are both switched simultaneously to 0, the inputs to both NOR gates will be at 0. The gates will both output 1, and X (the output of gate 1) will be propagated to X′ (the input of gate 2), while Q (the output of gate 2), will be fed back to Q′ (the input of gate 1). If the propagation times are identical, X′ and Q′ will become 1 at the same instant. Both gates will have inputs 10 and will output 0. X and Q become 0 again. The switching procedure then repeats itself and the output will oscillate between 0 and 1. In practice the delays will never be exactly the same. If X to X′ is shorter than Q to Q′ the output will remain at 0 (state ①). If not, state ② is entered and the output becomes 1. The final state of the circuit therefore will depend on the internal propagation delays.

The designer can avoid this flip-flop hazard by never allowing S and R both to be 1 at the same time. The problem associated with switching inputs from 11 to 00 will then never occur. If this restriction is enforced, the output, X, of gate 1 then becomes the inverse of the output of gate 2, i.e. $X = \overline{Q}$ for all allowable inputs.

If the flip-flop is regarded as a memory, SR = 11 represents the impossible request to set a single bit to both 1 and 0 at the same time.

The final truth table of an SRFF is:

S	R	Q_t	Q_{t+dt}
0	0	0	0
0	0	1	1
0	1	0	0
0	1	1	0
1	0	0	1
1	0	1	1
1	1	0	d
1	1	1	d

If SR = 11 is not allowed, the corresponding output becomes 'don't care'.

where Q_t is the present internal state (equal to the output before the circuit has operated) and Q_{t+dt} is the next state and output after the circuit has responded to the input data SR.

dt is the switching time of the circuit.

The transition table for the SRFF defines the inputs required to give a specific data transition from present state to next state ($Q_t \rightarrow Q_{t+dt}$). There are four possible transitions, 0→0, 0→1, 1→1 and 1→0. By comparing Q_t with Q_{t+dt} in the truth table, we see that the transition 0→0 occurs if SR = 00 or 01. S must therefore be 0

Table 4.3 Transition Table for an SRFF

S	R	$Q_t \rightarrow Q_{t+dt}$
0	d	0 → 0
1	0	0 → 1
d	0	1 → 1
0	1	1 → 0

but R can be 0 or 1, so R is a 'don't care'. The other transitions can be detected on the truth table and the input requirements are given in the SRFF Transition table (Table 4.3).

In this circuit the state and output transitions are identical.

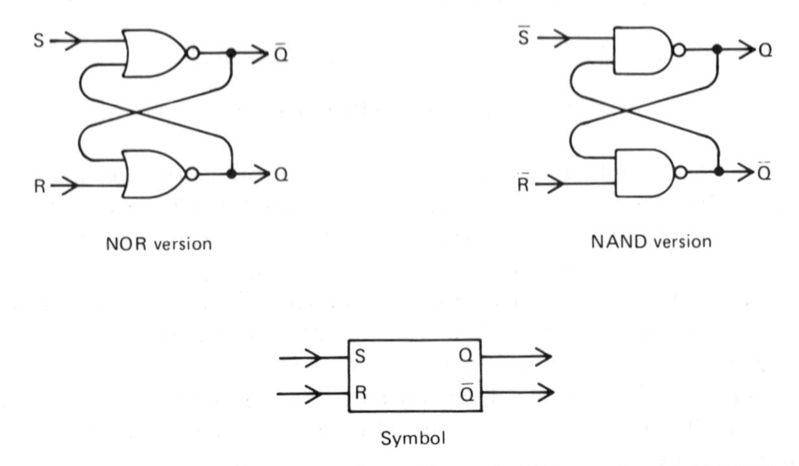

Fig. 4.3 Circuit diagrams of NAND and NOR versions of a SRFF.

Note the inverters on the inputs and outputs of the NAND circuit.

The circuit diagram of the SRFF is often drawn as a pair of cross-coupled gates (see Fig. 4.3). The NOR version may be converted to NAND by Boolean algebra. The SRFF is a simple sequential circuit having only one feedback loop. It is asychronous and operates solely on the input data S and R. No control data is required to determine when the circuit will switch. It operates immediately the inputs are applied. SRFFs form the basis of a family of bistables which have additional circuitry to overcome some of their practical limitations. Flip-flops can be regarded as the elementary building blocks of sequential systems.

Switch Debouncing using a Set-Reset Flip-Flop

One important and direct application of an SRFF is the debouncing of an electromechanical switch. A typical switch arrangement that produces a logic output F and its inverse $\bar{F}$ is shown in Fig. 4.4. The F and $\bar{F}$ terminals are connected to earth

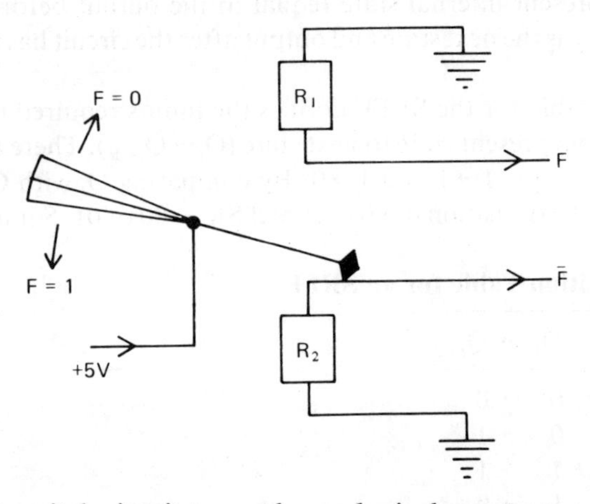

Fig. 4.4 A switch circuit to produce a logical constant and its inverse.

via resistors R_1 and R_2 and the toggle contact is at + 5 V. When the switch is in the F = 0 position, there is no voltage across R_1, F is at earth potential (Logical 0), 5 V appears across R_2 and $\overline{F}$ is at logical 1. While the switch is being operated, there will be a short period of time when the toggle is between the contacts of F and $\overline{F}$ and both outputs will be 0. When the toggle reaches F, it will become 1, but the toggle will physically bounce off the F contact, leaving the surface, momentarily causing F to be 0 again. It will then rebound an indeterminant number of times causing a string of 0s and 1s to be output before settling to F = 1. The bounce may be eliminated electronically, by connecting the switch output F to the S input, and $\overline{F}$ to the R input of an SRFF. The outputs Q and $\overline{Q}$ will then be the debounced versions of the switch outputs.

Bouncing is a mechanical effect that occurs in all simple switches. The electronics is sensitive to bounces over a very short period of time, typically 1 microsecond.

Worked Example 4.2

Show, using data waveforms, that an SRFF eliminates bounce from the switch arrangement in Fig. 4.3.

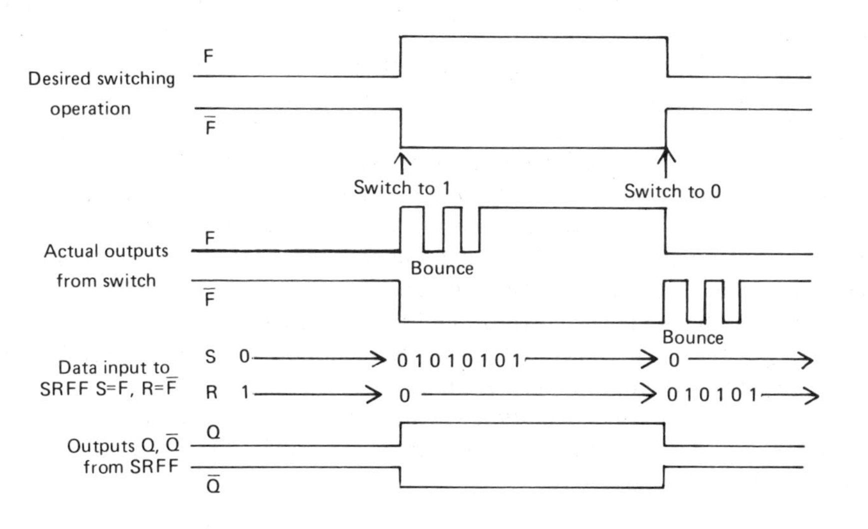

Q and $\overline{Q}$ give the desired bounce-free waveforms because the flip-flop is set (or reset) on the first touching of the contacts. The first bounce causes S = R = 0 and this does not change the outputs Q and $\overline{Q}$. The second time the contacts touch, the data will set (or reset) the flip-flop, but it has already been set (or reset) by the very first contact. There is consequently no change in Q and $\overline{Q}$ and the spurious 1s and 0s due to the bouncing, are eliminated.

Bouncing only occurs on the contacts that are 'making'. The 'breaking' contacts always give a clean pulse.

Refer to SRFF truth table.

Flip-flop is set when the switch goes to 1 and reset when switch goes to 0.

Exercise 4.1

Show by means of data waveforms that two cross-coupled NAND gates will eliminate the bounce on a switch which has its toggle contact earthed (logical 0) and dropper resistors connected to +5 V (logical 1).

The JK Flip-Flop

The input restrictions on the SRFF may be overcome if a JK flip-flop (JKFF) is used. This bistable consists of an SRFF with additional gating logic on the inputs.

Restriction. S = R = 1 is not allowed.

Note: J is ANDed with $\overline{Q}$ and drives S
K is ANDed with Q and drives R.

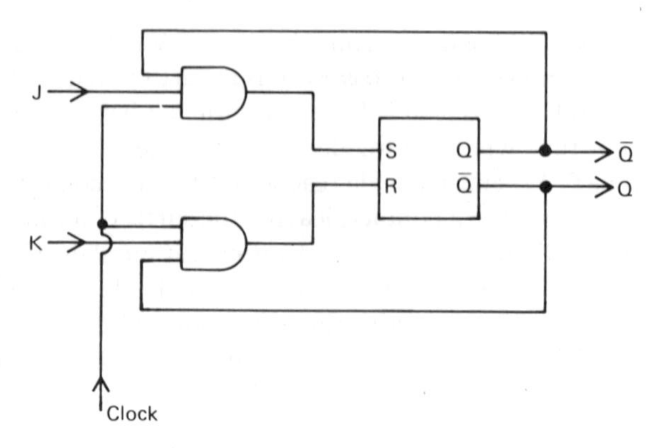

Fig. 4.5 Simple JK flip-flop.

The logic diagram of the JKFF is given in Fig. 4.5. The external inputs are J and K where J is ANDed with $\overline{Q}$, the inverse of the output of the SRFF, and drives S. K is ANDed with Q and drives R.

J and K have no literal meaning.

A third input, the clock, is common to both AND gates. Its function is to control the switching of the bistable and to synchronize the inputs. If the clock is at logical 0, both AND gate outputs to S and R will be 0 and the flip-flop will not therefore change state. The reader can evaluate the behaviour of the JKFF by referring to the SRFF truth table and calculating the logic values passing through the input logic, for all possible values of J and K. In particular, consider the case where both inputs are set to 1. If Q and the clock are at 1, the logic values transmitted through the AND gates are $S = 0$ and $R = 1$. This will reset the SRFF to $Q = 0$ and $\overline{Q} = 1$. These new outputs, when fed back via the input logic, will set SR to 10 and its outputs will become $Q = 1$, $\overline{Q} = 0$. Hence the output of the JKFF will oscillate continuously between 0 and 1 when both J and K are at 1.

This input condition is not allowed on an SRFF.

Although both J and K are 1, S and R can never both become 1 at the same time.

In the simple JKFF this oscillation will always happen, whereas with both inputs set to 1 on an SRFF, the final value of the output can be 0 or 1 depending on the internal delays within the circuit.

The Master–Slave Principle

Although the input logic in a simple JKFF eliminates some of the timing problems inherent in the SR device, its final state still cannot be determined when $J = K = 1$, as it depends on the length of time the circuit is allowed to oscillate. If the circuit was only allowed to switch once, then the behaviour could be completely specified. Single switching could be achieved by using a very short clock pulse, but this would give rise to technical difficulties. The alternative is to use a master - slave device.

A master–slave JKFF (Fig. 4.6) uses two SRFFs, together with input gating logic. The output of the second, or slave, SRFF is fed back to the input logic of the first SRFF (master). A most important feature is the inverter in the clock line. The clock is directly connected to the master but passes through an inverter before entering the slave. When the clock is high, the inputs switch the master and the feedback from Q_2, and $\overline{Q}_2$ prevents S_1 and R_1 from being simultaneously 1. The slave inputs S_2 and R_2 remain at 0 owing to the inverter in the clock line inhibiting the output of gates 3 and 4. As the clock switches to 0, data can be input to the slave

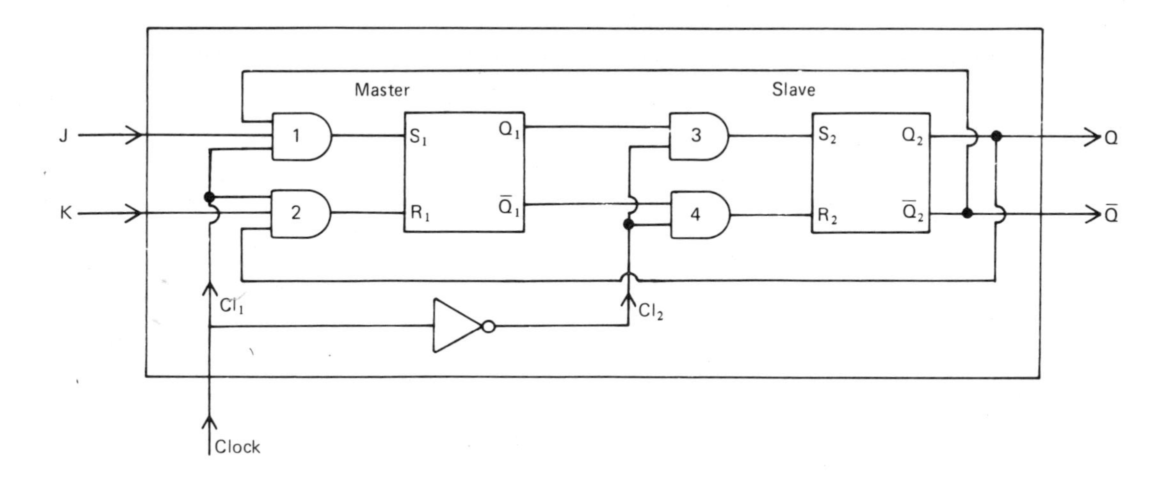

Fig. 4.6 A master-slave JKFF.

as its clock, Cl_2, now becomes 1. The slave inputs are the outputs of the master, which remain constant because Cl_1 is 0 and no new data can enter the master. The output of the slave is determined by the state of the master which is governed by the values of J and K the instant the clock switches from 1 to 0. The master–slave system can therefore be considered as a JKFF which is switched by the data on the inputs at the instant a falling edge occurs on the clock line. Only one change of state Q can occur for each clock pulse.

Master does not change because $S_1 = R_1 = 0$.

The truth table for a master–slave JKFF is:

J	K	Q_t	Q_{t+1}	Effect on output
0	0	0	0	no change
0	0	1	1	
0	1	0	0	reset to 0
0	1	1	0	
1	0	0	1	set to 1
1	0	1	1	
1	1	0	1	change state
1	1	1	0	

Q_t is often called the PRESENT state and Q_{t+1} the NEXT state.

t is measured in clock pulses.

where J and K are the external inputs, Q_t is the output (and internal state) before, and Q_{t+1} is the output after the clock edge has occurred.

The transition table which gives the required inputs for each output transition is:

J	K	$Q_t \rightarrow Q_{t+1}$
0	d	$0 \rightarrow 0$
1	d	$0 \rightarrow 1$
d	0	$1 \rightarrow 1$
d	1	$1 \rightarrow 0$

Almost all commercially available JKFFs are of the master–slave type and operate on the falling edge of the clock. If any circuit contains JKFFs they should be assumed to be the master–slave type.

Worked Example 4.3 The following data is applied to J, K and the clock of a master – slave JKFF. If its initial state is Q = 0, what is the output waveform?

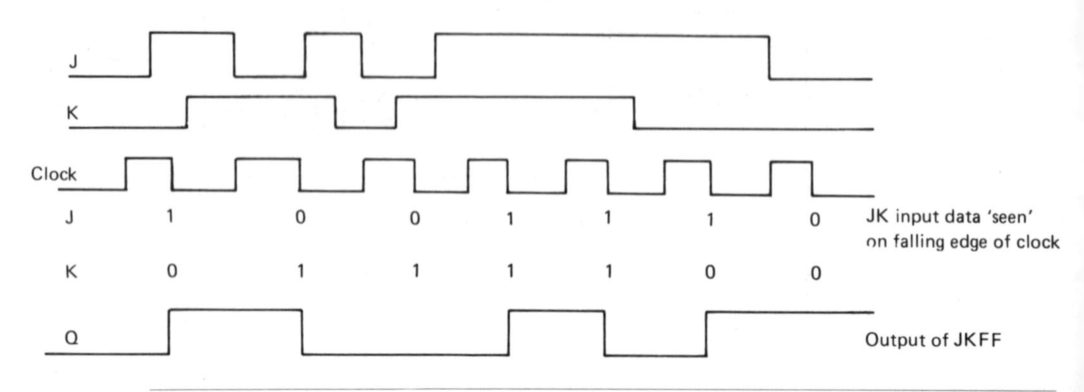

Exercise 4.2 Design a master-slave JK flip-flop using NOR gates only.

The Trigger Flip-Flop

Sometimes called a toggle.

Two types of trigger flip-flops (TFF) may be designed by restricting the input conditions on a JKFF. In the asynchronous TFF the inputs to the internal JKFFs are connected permanently to 1, and data is input via the clock line, as shown in Fig. 4.7. The output changes each time a falling edge occurs on T. There will be one change on Q either (0→1 or 1→0) for each complete pulse on T. Typical input and output waveforms are

The internal JKFF must be a master–slave device.

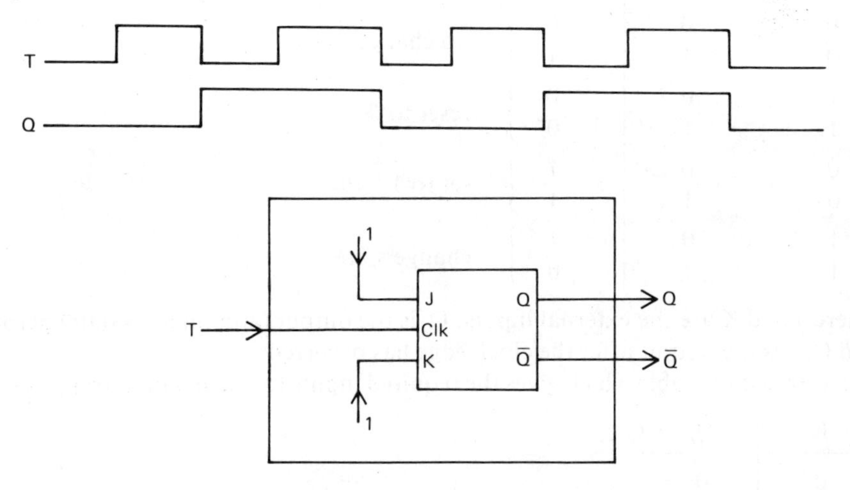

Fig. 4.7 An asynchronous TFF.

The output Q contains exactly half the number of pulses in the input T. As the asynchronous trigger divides the input by 2, it may therefore be used to build up binary counting circuits. Let the output of an asynchronous TFF drive another one as shown in Fig. 4.8. The waveforms on T_0, Q_0 and Q_1 will be

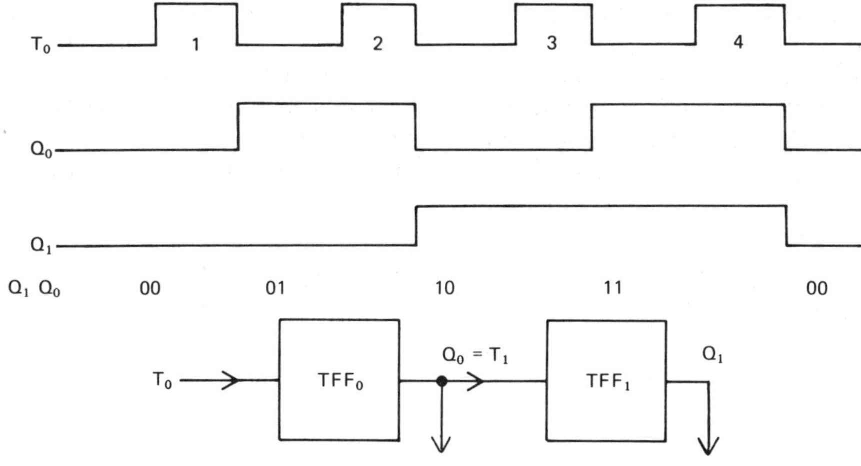

Fig. 4.8 A 2-bit pure binary counter.

The binary value of Q_1 Q_0 immediately after a pulse on T_0, is numerically equal to the number of pulses that have occurred on T_0. The system is a 2-bit pure binary counter, and its range can quite easily be extended by adding on extra TFFs.

Exercise 4.3

Verify that the binary counter in Fig. 4.8 will count down if either Q is propagated between flip-flops and $\overline{Q}$ used for output, or Q is output and $\overline{Q}$ is propagated. What happens if $\overline{Q}$ is both output and propagated?

The Synchronous Trigger Flip-Flop

A synchronous TFF can be constructed from a master–slave JKFF, by connecting J and K together to form the trigger (T) data input. The clock input receives the control information which determines when the device switches.

The diagram of a JKFF wired up as a synchronous trigger is shown in Fig. 4.9. As J and K are connected together, the input values that can be applied to the internal JKFF are restricted. The truth table for the synchronous TFF will comprise those entries from the JK table where J and K have the same value.

The transition table for the synchronous trigger is:

T	$Q_t \rightarrow Q_{t+1}$
0	$0 \rightarrow 0$
1	$0 \rightarrow 1$
0	$1 \rightarrow 1$
1	$1 \rightarrow 0$

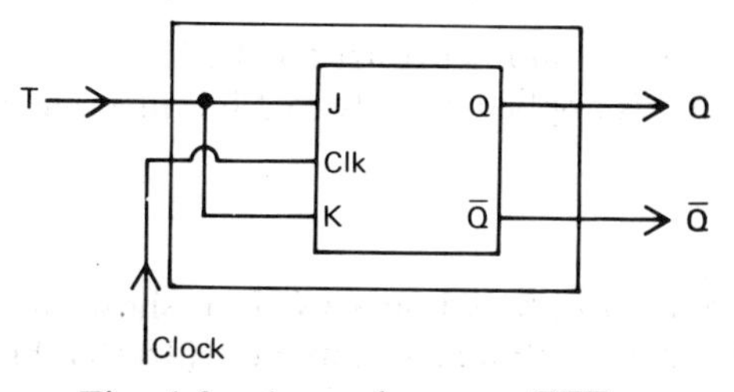

Fig. 4.9 A synchronous TFF.

To change the output state (0 → 1) or (1 → 0) the T input must be 1. If the output is to remain constant (1 → 1) or (0 → 0) then T has to be 0. The transition $Q_t \rightarrow Q_{t+1}$ takes place when a falling edge occurs on the clock input.

A synchronous TFF cannot be used to count a sequence of pulses applied to its T input. Although the output behaviour is determined by the value of the data at T, the timing of the switching is controlled by the clock. Consider the waveforms T and clock applied to a synchronous trigger.

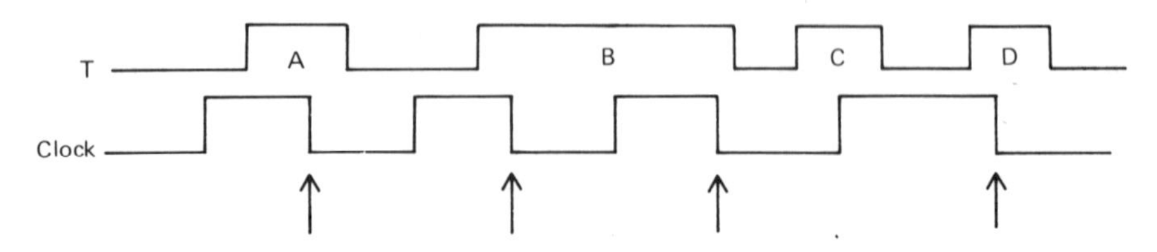

Although the data is continuous, the flip-flop only 'sees' values of T on the falling edges of the clock (indicated by arrows). The flip-flop will therefore completely miss input pulse C and respond twice to pulse B.

Now consider a clock frequency that is much faster than the data input:

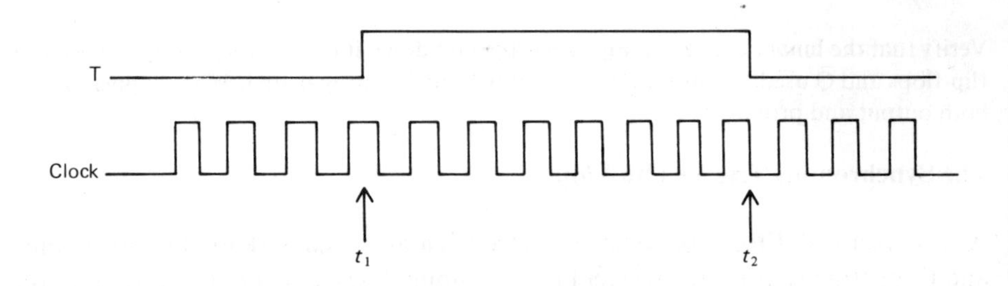

The flip-flop will not change state before time t_1 as the input T is 0 on each clock falling edge. At t_1 the data input T goes high and the flip-flop output will change state on every falling clock edge, between times t_1 and t_2. After t_2 the input becomes 0 and the output remains constant.

When T = 1 and $Q_t = 0$, $Q_{t+1} = 1$, $Q_{t+2} = 0$ etc.

If the frequency of the clock is known, the number of switching operations in the flip-flop will depend on the duration of the input pulse. Synchronous TFFs can therefore be used in the building block of timing circuits. These will be examined in detail in Chapter 5.

The Delay Flip-flop

The delay flip-flop (DFF) is another restricted JKFF. It is synchronous and has a single data input D that is applied directly to J but inverted before going to K.

Hence $\quad J = D$
and $K = \overline{D}$

The block diagram of a DFF, based on a JKFF, is shown in Fig. 4.10.

The transition table can be obtained from the JK truth table. Valid D inputs are when $J = \overline{K}$.

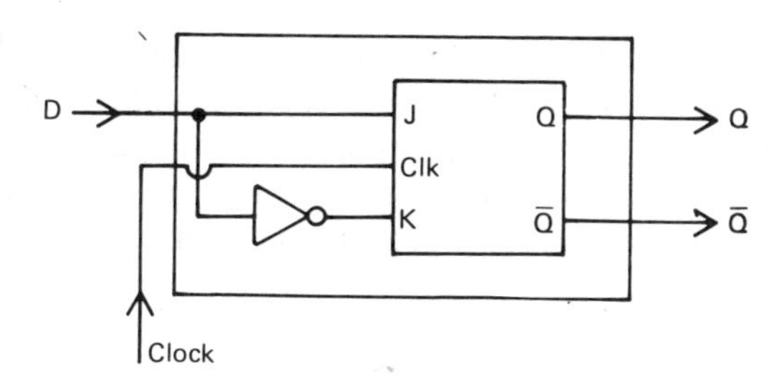

Fig. 4.10 A delay flip-flop.

D	$Q_t \rightarrow Q_{t+1}$
0	$0 \rightarrow 0$
1	$0 \rightarrow 1$
1	$1 \rightarrow 1$
0	$1 \rightarrow 0$

J	K	Q_t	Q_{t+1}	
0	0	0	0	
0	0	1	1	
0	1	0	0	Valid D = 0
0	1	1	0	Valid D = 0
1	0	0	1	Valid D = 1
1	0	1	1	Valid D = 1
1	1	0	1	
1	1	1	0	

The device is called a DFF because the next output Q_{t+1} always takes the value of the input D, regardless of the present state Q_t. It may be used to introduce a controlled delay into a logic system, the delay being dependent on the external clock frequency. If the clock is stopped the DFF will store its last input.

Shift Registers

A shift register may be formed by connecting DFFs in series. The clocks of the individual DFFs are all connected together as shown in Fig. 4.11. After every clock pulse the data stored in each flip-flop is shifted into its neighbour. Data may be input, one bit at a time to DFFo and propagated through the shift register to the final output from DFF*n*. Such devices are known as serial-in, serial-out (SISO) shift registers and can have a capacity of up to 2k-bits.

Compare the shift register with the counter which is asynchronous TFFs in series.

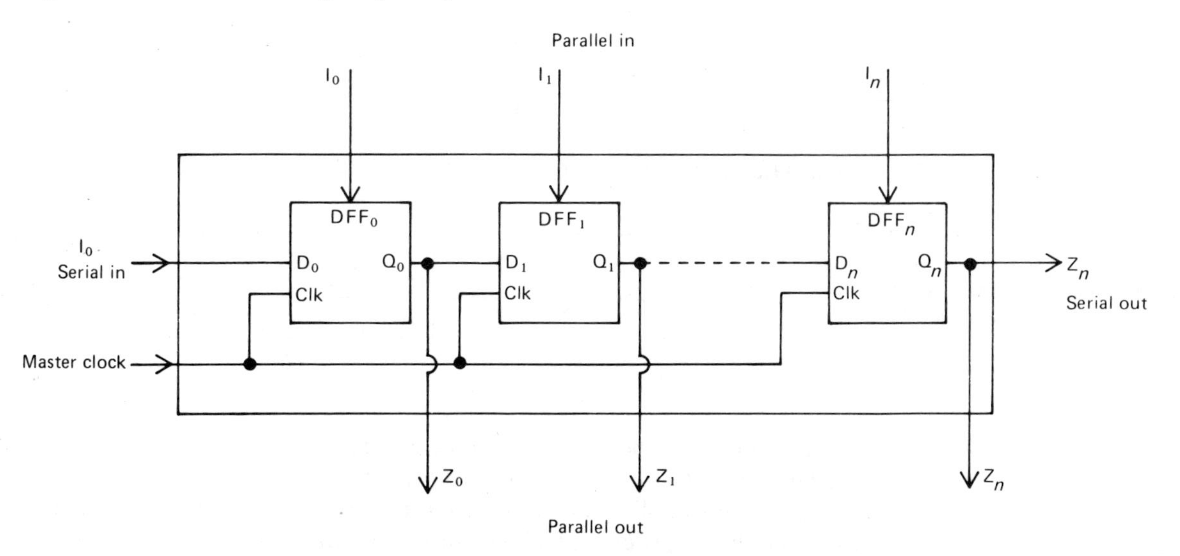

Fig. 4.11 A shift register.

If preset and clear controls are used on each DFF, data can be loaded into the shift register in parallel. Connections from each individual flip-flop output will create a parallel data output path, giving a parallel-in parallel-out (PIPO) shift register. Owing to the large number of input and output lines required by a PIPO, they are restricted to at most 16 bits, in their single chip integrated form.

Shift Register Applications

'Clocked' means one clock pulse has been applied.

Parallel-to-serial converter. Parallel data occurring on several lines may be converted into a string of serial information on a single line by using a parallel-in, serial-out shift register. The data is loaded into the shift register via the parallel inputs. The register is clocked n times, where n is the number of bits in the word. Each clock pulse will cause 1 bit to be output on the serial output line Z_n.

Worked Example 4.4

Use a shift register to convert a 4-bit parallel word 1011 into a serial stream of data.

A 4-bit parallel-in serial-out shift register is needed. The word 1011 is applied to the parallel inputs

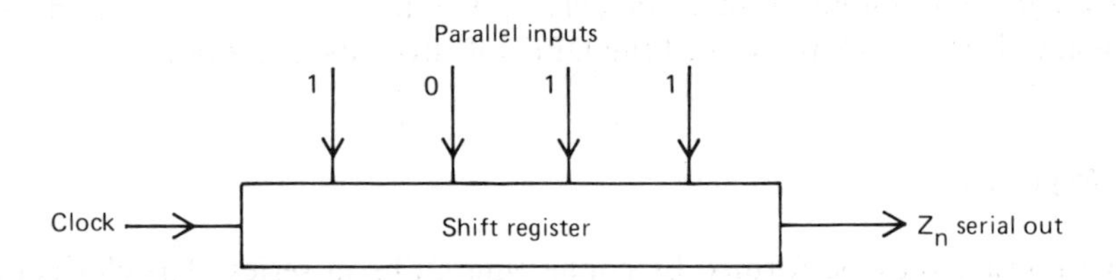

The states and outputs Z_n are as follows:

Clock pulse	State	Z_n
1	1011	1
2	0101	1
3	0010	0
4	0001	1

The data on Z_n is therefore

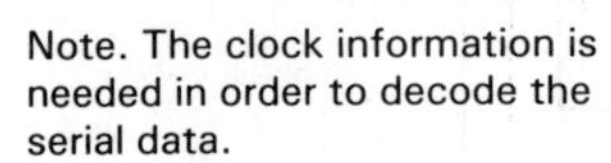
Note. The clock information is needed in order to decode the serial data.

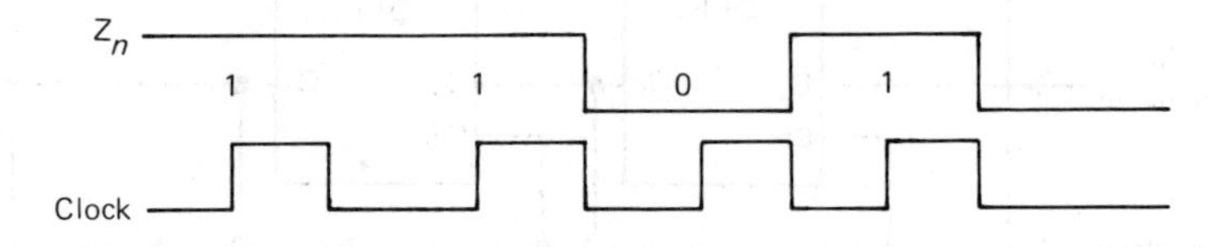

The connections between the data lines and the shift register will determine whether the least or most significant bit is output first on Z_n.

One common application of a parallel-to-serial converter is in a multiplier circuit where the shift register controls the successive addition.

Serial-to-parallel converter. By using the serial input and the parallel outputs of a

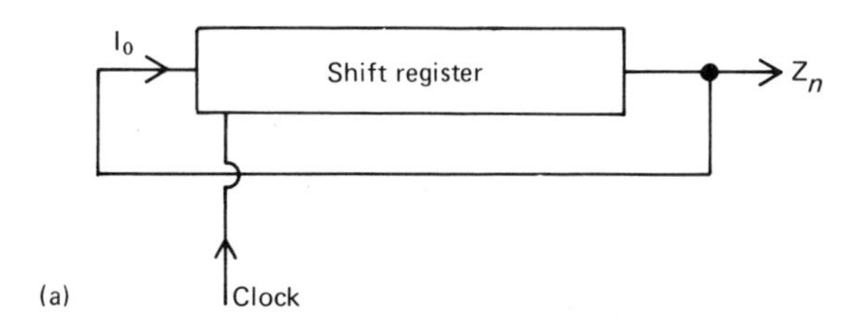

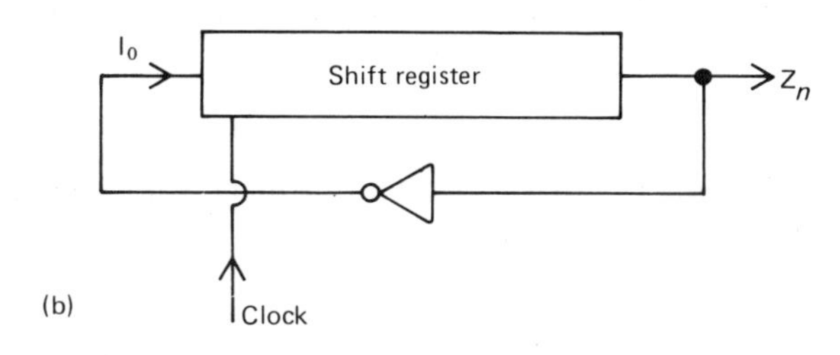

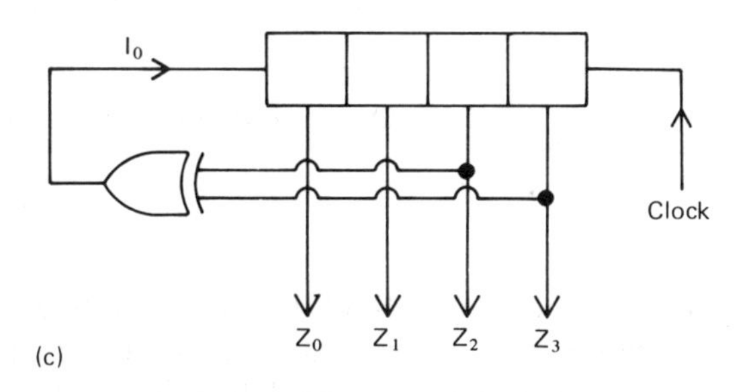

Fig. 4.12 Feedback Shift Registers. (a) Normal feedback. (b) Inverted feedback. (c) EXCLUSIVE- OR feedback.

shift register, a data stream on a serial line can be distributed over the parallel outputs. This operation is the dual of the parallel to serial system.

Shift registers with feedback. If feedback is applied to a shift register, a repeatable sequence of states may be generated. The serial output or its inverse may be fed back to the serial input or alternatively the input could be a function of two or more of the parallel outputs.

By feeding back the serial output to the input (Fig. 4.12), a shift register can perform a parallel to serial conversion without losing its data. If the register's initial state is 1011 the states after each shift and feedback will be

Shift 0 is the initial state.

Shift	State
0	1011
1	1101
2	1110
3	0111
4	1011

After 4 shifts, all the state bits have been output on the serial port, but the original data is still in the register. In general, the data in an n-bit register with normal feedback, is restored after n shifts.

An inverted feedback shift register has the inverse of its output fed back (Fig. 4.12b).

If an n-bit word in an inverted feedback shift register is shifted n times, the original word is inverted. A further n shifts restores the original word.

Worked Example 4.5 Determine the state sequence of an inverted feedback shift register whose initial state $Q_3\ Q_2\ Q_1\ Q_0$ is 1011 and $I_0 = Q_3$.

	Q_3	Q_2	Q_1	Q_0	
Initial state	1	0	1	1	
1st shift	0	1	1	0	$Q_0 \rightarrow Q_1$, $Q_1 \rightarrow Q_2$, $Q_2 \rightarrow Q_3$ but $\overline{Q}_3 \rightarrow Q_0$
2nd shift	1	1	0	1	
3rd shift	1	0	1	0	
4th shift	0	1	0	0	Inverse of initial state
5th shift	1	0	0	1	
6th shift	0	0	1	0	
7th shift	0	1	0	1	
8th shift	1	0	1	1	Original data restored after $2n$ shifts

If the feedback function is the EXCLUSIVE-OR of two parallel output bits as in Fig. 4.12(c) the shift register will generate a sequence of apparently random numbers. The sequence can be repeated provided the shift register always starts in the same state and the feedback function is not changed. The random number sequence contains a maximum of $2^n - 1$ states, where n is the number of bits in the register. Linear feedback shift registers can be used in the coding of information for security purposes, when for example confidential data is transmitted over a public communications channel. The data is encoded with a random number generator and may be decoded if the feedback function and the starting state of the shift register are known.

Worked Example 4.6 The message 'This is secret' is stored letter by letter in addresses 1 to 15 of a memory. Devise a feedback shift register to encode the message before transmission and to decode it at the receiver.

Message in Transmitter

Memory address	Contents	Memory address	Contents
1	T	9	S
2	H	10	E
3	I	11	C
4	S	12	R
5		13	E
6	I	14	T
7	S	15	.
8			

FSR is 'feedback shift register'. Shift register stores word $Q_3 Q_2 Q_1 Q_0$ and I_0 is its serial input.

Now consider a 4-bit FSR with $I_0 = Q_2 \oplus Q_3$ and starting state 1011. Its number sequence is

	Q_3	Q_2	Q_1	Q_0	Decimal
Initial state	1	0	1	1	11
1st shift	0	1	1	1	7
	1	1	1	1	15
	1	1	1	0	14
	1	1	0	0	12
	1	0	0	0	8
	0	0	0	1	1
	0	0	1	0	2
	0	1	0	0	4
	1	0	0	1	9
	0	0	1	1	3
	0	1	1	0	6
	1	1	0	1	13
	1	0	1	0	10
	0	1	0	1	5
(Repeat sequence)	1	0	1	1	11

After each shift $Q_2 \rightarrow Q_3$, $Q_1 \rightarrow Q_2$, $Q_0 \rightarrow Q_1$, but $Q_3 \oplus Q_2 \rightarrow Q_0$.

The random number sequence is used to address the transmitter memory and the message transmitted will be

CS.TR THSSIIEE

At the receiver the message can be decoded by using the same random sequence. The complete set of numbers DOES NOT have to be transmitted. It can be generated in the receiver with another FSR, provided both the feedback function ($I_0 = Q_3 \oplus Q_2$) and the initial state 1011 are known. The letters received will be written into a memory whose addresses are determined by the FSR.

Letter received	Address	Letter received	Address
C	11	S	4
S	7	S	9
.	15	I	3
T	14	I	6
R	12	E	13
	8	E	10
T	1		5
H	2		

The decoded message in the receiver memory will be

Memory address	Contents	Memory address	Contents
1	T	9	S
2	H	10	E
3	I	11	C
4	S	12	R
5		13	E
6	I	14	T
7	S	15	.
8			

Summary

A sequential circuit is formed by applying feedback to a combinational logic system. Feedback creates additional internal inputs (or states) that cannot be controlled directly by the external inputs. They are dependent on the previous inputs and state. The latter is dependent on earlier inputs and states, and so on. Hence a sequential circuit's behaviour is a function of all its previous inputs.

The stability of a sequential circuit can be summarized on a flow table. For a given value of external inputs, the circuit will always switch into a stable state unless a recurring cycle of unstable states are entered, causing an oscillation.

The SRFF is a sequential circuit with a single feedback loop. Its circuit is a pair of cross-coupled gates (either NAND or NOR). It is a memory device, but has inherent timing problems. The timing problems in the SRFF are overcome in the master–slave JKFF.

If J and K are permanently set to 1 and data entered via the clock input, we have the asynchronous trigger which forms the basis of many counting circuits. With J and K connected together to form a common data input and a separate clock, we have the synchronous trigger that can be used in timing circuits. Finally, if an inverter is then inserted between J and K such that the data input goes to J and its inverse to K, the result is a DFF.

DFFs connected in series form a shift register which has wide applications when data has to be converted between parallel and serial forms. A shift register with feedback can be used to retain information, to invert binary words and to create sequences of random numbers.

Problems

4.1 Analyse the following sequential circuit and summarize its behaviour on a flow table.

$$F = A + \overline{\overline{B}.F'}$$

where F is fed back to form internal input F′.

4.2 A SRFF is driven from a 2-bit binary counter $b_1\,b_0$. Output b_1 is ANDed with $b_1 \oplus b_0$ and drives R. b_0 is also ANDed with $b_1 \oplus b_0$ and drives S. Evaluate the output from the SRFF as the counter progresses through its 4 states starting at 00. (The flip-flop is initially reset to 0).

4.3 A JKFF is driven from a 2-bit binary counter with b_1 connected to J and b_0 connected to K. The flip-flop and counter have a common clock and the flip-flop is initially reset to 0. What is its output sequence as the counter is incremented from 00? Comment on any possible timing problems.

4.4 Design a master–slave JKFF using only NAND gates. What is the minimum number of gates required?

4.5 A sequential system comprises two feedback logic functions:

$$F_1 = AB\overline{F_1}' + F_1'\overline{F_2}'$$
$$\text{and} \quad F_2 = (A+B+F_2')(\overline{A}+\overline{B}+F_2')$$

where $F_1 \rightarrow F_1'$ and $F_2 \rightarrow F_2'$
Under what input conditions will the system oscillate?

4.6 A manufacturing error results in a batch of flip-flops with cross-coupled NOR/NAND gates as shown below:

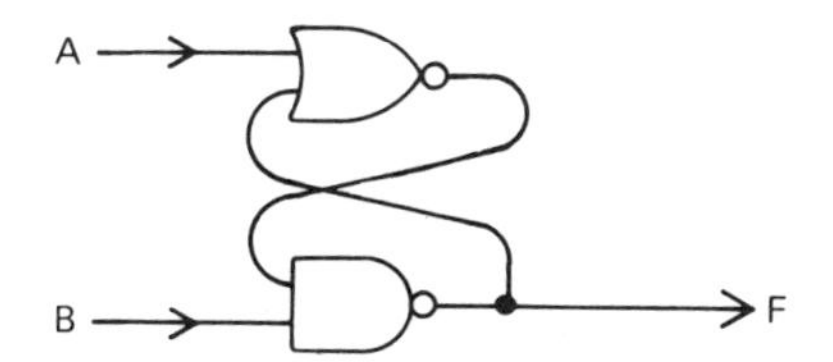

Show that this device *cannot* be reset via its input terminals AB.

4.7 A circuit comprises a synchronous trigger TFF0, driven by input I. Its output Q_0 drives a delay flip-flop DFF1 whose output, Q_1 in turn drives another synchronous trigger TFF2. Determine the sequence of outputs on $Q_2Q_1Q_0$, starting from 000, that occur when I is set to 1 and the circuit clocked.

4.8 Design a logic system using 2 SRFFs and any necessary AND/OR/NOT logic to control 2 machines. Each machine is turned on by a flip-flop if its output is 1 and off if 0. The logic system operates from 3 push buttons A, B and C. When A is pressed, both machines are switched on. If both machines are on and B is pressed, machine 1 must be turned off. Any machine that is on when C is pressed, is turned off. Also, devise a digital input circuit to prevent the input data from being transferred to the controller if 2 or more buttons are pressed simultaneously.

4.9 Determine the count sequence, starting at 000, when 3 synchronous triggers are connected in series, the input set to 1 and the system clocked.

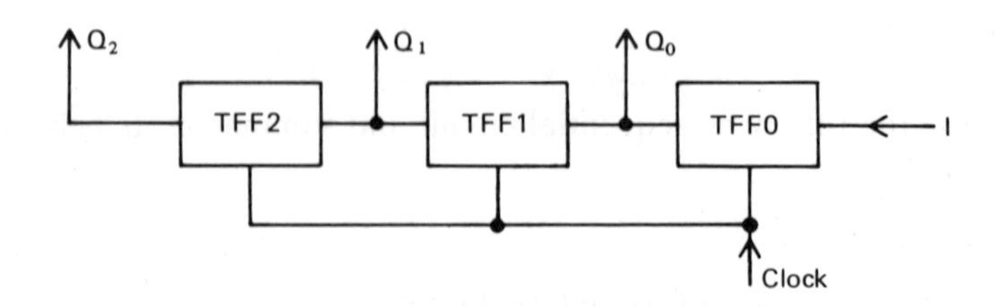

Compare and contrast this counter with a 3-bit version of the asynchronous counter given in Fig. 4.8.

4.10 A 4-bit shift register $Q_3Q_2Q_1Q_0$ has a feedback function

$$I = \overline{Q_3.(Q_2 \oplus Q_1)}$$

where I is the serial input. What is its output sequence if its starting state is 0000?

Design of Sequential Logic Circuit

5

Objectives

- ☐ To identify the parameters and structure of a general sequential circuit.
- ☐ To design asynchronous counting circuits using trigger flip-flops and to investigate the use of output encoders to enable count sequences other than pure binary, to be implemented.
- ☐ To examine the limitations of asynchronous circuits and the use of synchronous circuits to overcome these practical problems.
- ☐ To evaluate the JKFF as a building block for synchronous systems and the design of its driving logic.
- ☐ To specify sequential systems by means of present/next state tables.
- ☐ To apply state transition diagrams to sequential logic design.
- ☐ To simplify sequential systems by state minimization.

In Chapter 4, we saw that a sequential circuit was basically a combinational logic circuit with feedback, and a range of simple feedback circuits called flip-flops could be devised. This chapter will concentrate on the design of logic systems using these discrete building blocks. As every sequential circuit contains a large amount of combinational logic, the methods and techniques outlined in Chapter 3 will be required when designing sequential systems.

The Parameters of a Sequential Circuit

In any sequential logic system three distinct sets of data and two logic functions can be identified. The input data is represented by the set **I** and the output by **Z**. The sequential system generates internal data or states, which form additional internal inputs to the logic functions. These internal states form the data set **Q**. The logic function that generates the internal states has the symbol δ, and this is implemented by a combinational logic circuit operating on the external inputs and the internal state, (the outputs of δ), fed back to form internal inputs. The output of δ is the next state of the system and the data on the internal inputs is the present state.

The internal state is often simply referred to as the 'state'. δ is pronounced 'delta'

Note. The output of δ (the next state) when fed back becomes the internal inputs (the new present state).

The next state logic system is responsible for the circuit's sequential properties that arise from the feedback connections. This part of the circuit can be partitioned into flip-flops — the standard circuits with a single feedback connection which were examined in the previous chapter.

The second logic function within a general sequential system is ω, a combinational logic circuit which generates the output **Z**. Its inputs are the external inputs **I** and the state **Q**. The general form of a sequential logic system is given in Fig. 5.1.

ω is the Greek letter omega.

The output logic can be obtained from the truth table which specifies the outputs for all possible combinations of inputs and internal states. In some sequential

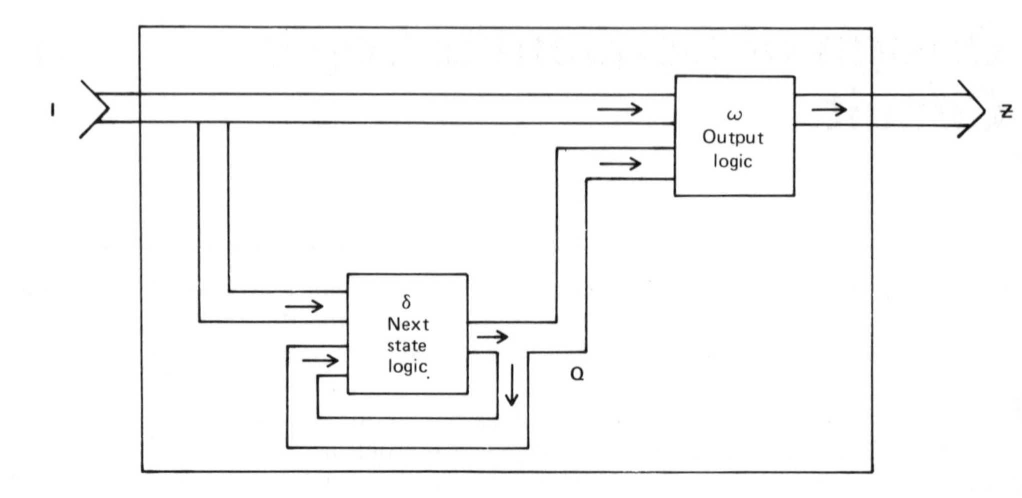

Fig. 5.1 A general sequential system.

circuits there is no need for output logic, as the outputs are identical to the internal states. Counters and code generators do not require any output logic, unlike 'one in n' and sequence defectors where the output is a function of the inputs and the state of the system.

See Worked Example 5.5.

The feedback applied to the next state logic is crucial to sequential behaviour. The logic outputs that represent the internal state of the system are fed back to provide internal inputs. This enables the next state logic to give different responses to identical external inputs applied to the circuit at different times in a sequence of inputs. Suppose a logic system has two inputs I_1 and I_2, and two state variables Q_1 and Q_2. Let the external inputs be 00 and the state also 00. The four inputs to the next state logic will be $I_1\ I_2\ Q_1\ Q_2 = 0000$ and an output will be generated. Let the new output, which is the next state of the system, be 01. This is fed back and the input data to the next state logic becomes 0001. The logic designer can ensure that a new output is generated from this data. Suppose the new output is 11. It is now possible to distinguish between the first and second occurrences of 00 on the inputs. When 00 is first applied, the internal state becomes 01. A subsequent input of 00 switches the state to 11. Detection of the different occurrences of input data, in a sequence, is possible even though the data is numerically equal to earlier or later inputs. Hence the system is sensitive to the sequence of the inputs.

This would only be attempted in a synchronous system where the inputs are synchronised with a clock.

The designer of the next state logic has to be able to identify the internal states of the system and then design hardware to generate them. The hardware can be implemented in one of three forms. It can be based on either

(i) Combinational logic with feedback or
(ii) Standard flip-flops or
(iii) Semiconductor memory.

Sequential systems comprising combinational logic with feedback will not be examined in great depth here. Although these circuits can operate at very high speeds, they are prone to timing problems. The logic is asynchronous and any unequal delays in the feedback paths can cause the system to malfunction.

Feedback combinational circuits are sometimes called primitive circuits. For fuller discussion refer to Lewin, D. *Logical Design of Switching Circuits*, 2nd edn (Van Nostrand Reinhold, 1974).

The use of VLSI memory devices as building blocks for sequential logic systems will be examined in Chapter 6. These devices are suitable for larger systems and represent an extension of the principles pertaining to designs based on standard flip-flops which are, in effect, 1-bit memories. The remainder of this chapter will be devoted to the design of sequential logic using discrete flip-flops.

VLSI: Very Large Scale Integration.

Asynchronous Binary and Non-binary Counters

In Chapter 4 it was found that an asynchronous TFF would divide an input pulse train by 2 so that the output contained only half the number of pulses in the input. A binary counter can be formed by cascading several TFFs as shown in Fig. 5.2. This is in effect, δ, the next state logic for the counter.

Take care not to confuse 'asynchronous circuit' with 'a synchronous circuit' — they sound the same.

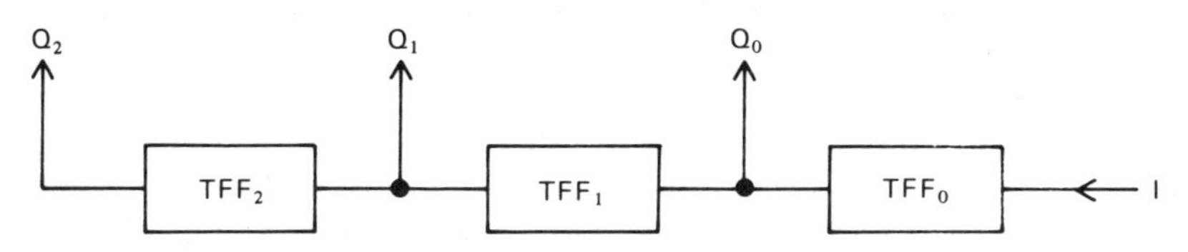

Fig. 5.2 An asynchronous binary counter.

A counter is a sequential logic circuit. A differential output (Q_2 Q_1 Q_0) equal to the value of the count, is required each time the input I is set to 1. The circuit in Fig. 5.2 does not depart from the general concept of the feedback circuit given in Fig. 5.1. The external input together with the internal state variables drives the next state logic which consists of 3 TFFs. The next state logic can be re-drawn as a feedback circuit as shown in Fig. 5.3.

Asynchronous TFFs switch on the falling edge of the inputs. The counter will be insensitive to the length of time the input is held at 1.

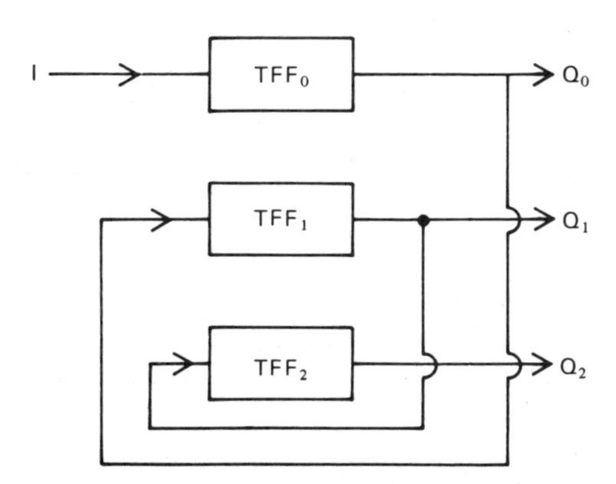

Fig. 5.3 An asynchronous binary counter arranged as a feedback circuit.

This is a very simple sequential circuit. There is no output logic and the feedback is connected directly to the terminals of the flip-flops. In more complicated systems the inputs to the flip-flops will be functions of both input and internal state variables. These functions are known as the 'driving logic'. A general block diagram of next state logic with driving functions is shown in Fig. 5.4.

The 'driving logic' is sometimes called the 'excitation' or 'steering logic'.

If a non-binary count is required, driving logic is needed to control the flip-flops, and its design is illustrated in the following Worked Example.

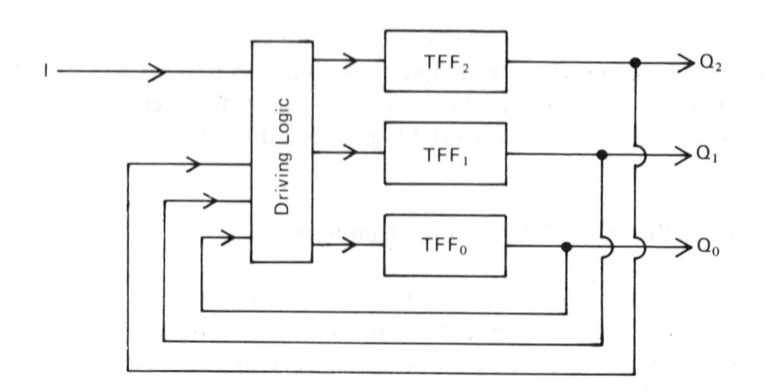

Fig. 5.4 Next state logic partitioned into flip-flops with driving functions.

Worked Example 5.1

Design an asychronous counter that, on the falling edge of each input pulse, progresses through the sequence 000, 010, 001, 011, 100, and repeats from 000.

Each state variable change must be identified, whether 0 to 1 or 1 to 0.

If asychronous TFFs are used, the count sequence must be examined in order to identify falling edges which can then be used to switch the flip-flops.

Q_2	Q_1	Q_0	State
0	0*	0	a
0	1*	0*	b
0	0*	1	c
0*	1*	1*	d
1*	0	0	e
0	0	0	a (repeated)

A falling edge must occur on the input I before every change in the count.

Each change of state of flip-flop output is denoted with a *. It can be seen that the input I cannot be used to switch any of the flip-flops, as none of the output bits change state on every input pulse. Q_0 changes its value as the count progresses from state b to c and from d to e. The falling edge on Q_1 occurring after states b

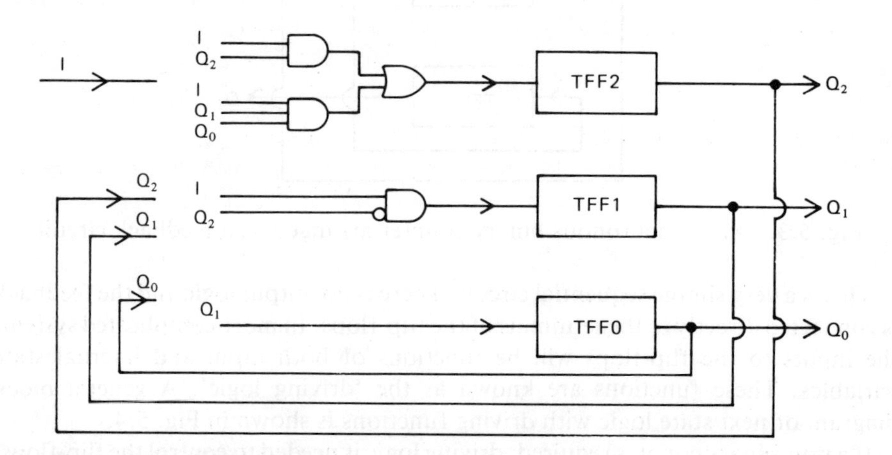

Fig. 5.5 Logic diagram of the non-binary counter in Worked Example 5.1.

and d can be used to switch TFF0. On further examination of the state sequence the following functions may be identified:

$$T_1 = I\overline{Q}_2$$

which will switch the second flip-flop giving bit Q_1 of the count and

$$T_2 = I\,(Q_1\,Q_0 + Q_2)$$

I must be included in the driving functions to prevent the counter from halting in states a or e.

which will switch the third flip-flop producing the most significant bit of the count.

The complete logic diagram for the system is given in Fig. 5.5

Exercise 5.1

Design an asynchronous counter using TFFs that will count in BCD. The counter must comprise four bits and the count sequence, starting at 0000, proceeds to 1001 and then returns to 0000 and repeats the sequence.

Practical Limitations of Asynchronous Counters

The design procedure when using edge-triggered flip-flops involves the identification of functions that will produce falling edges at points in the count sequence which coincide with the changing of one or more bits in the count. The edges can then be used to switch the flip-flops.

Two problems arise from this approach:

(i) It may not be possible to identify functions to give the desired flip-flop switching sequence.
(ii) The count may halt in certain states if the switching of two or more bits are inter-dependent.

Problem (ii) can be illustrated as follows. Suppose a count sequence required a 2-bit counter Q_1Q_0 to change from 11 to 00. The falling edge on Q_1 appears to be suitable to switch flip-flop T_0 and likewise $T_1 = Q_0$. In practice, however, each flip-flop would be waiting for the other to change and if there was no other data applied to the trigger inputs, the counter would halt in state 11.

As a result of these limitations, it is not possible to design certain non-binary sequence counters using asynchronous triggers and driving logic.

An Asynchronous Counter with Output Logic

The general form of a sequential circuit (Fig. 4.1) contains two logic systems, δ, the next state logic and ω, the output logic. However the counters in the previous section of this chapter did not require any output logic as the outputs were identical to the internal states. If a particular count sequence cannot be achieved using only next state logic because of the problems of identifying suitable switching functions and their possible inter-dependence, a pure binary counter with output logic has to be used. The output logic then encodes the internal states into the desired count sequence.

Worked Example 5.2

Design the 3-bit counter that will count a string of randomly occurring pulses in Gray code.

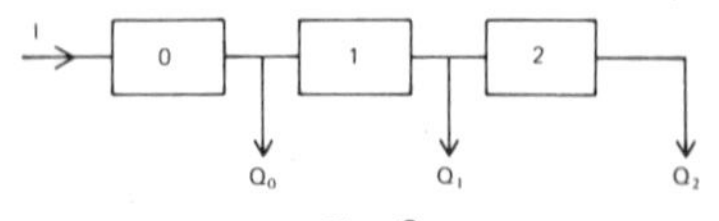

Fig. 2

The pulses can be counted in pure binary by cascading 3 TFFs (see Fig. 5.2). Output logic is required to convert the binary to Gray code according to the following truth table:

Q			Z		
Q_2	Q_1	Q_0	Z_2	Z_1	Z_0
0	0	0	0	0	0
0	0	1	0	0	1
0	1	0	0	1	1
0	1	1	0	1	0
1	0	0	1	1	0
1	0	1	1	1	1
1	1	0	1	0	1
1	1	1	1	0	0

Z_2, Z_1, and Z_0, are all functions of $Q_2\,Q_1\,Q_0$ and can be minimized on K-maps.

The EXCLUSIVE OR function is defined as $Q_2 \oplus Q_1 = Q_2\,\bar{Q}_1 + \bar{Q}_2\,Q_1$. $Z_2 = Q_2$ can be observed on the truth table.

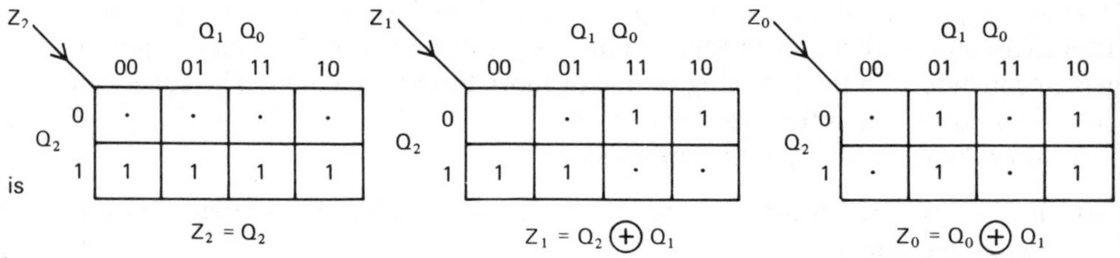

The complete logic diagram for the system is given in Fig. 5.6.

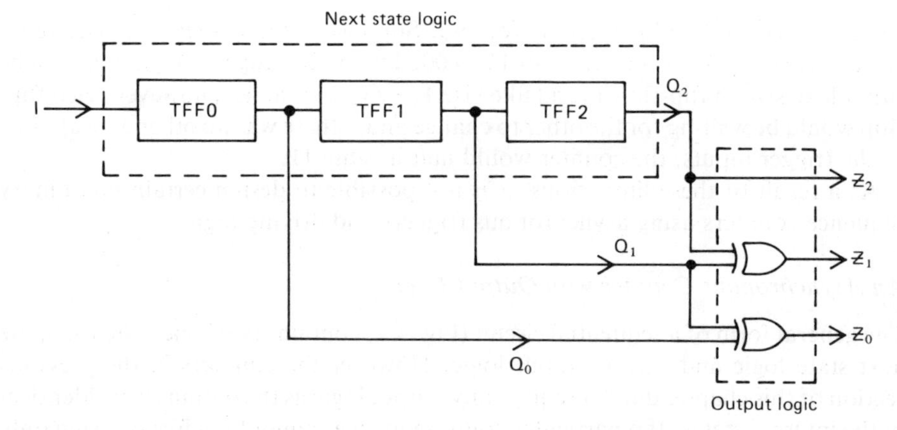

Fig. 5.6 A Gray code counter based on a pure binary counter with output logic.

The effects of next state propagation delays are eliminated in a synchronous system.

A counter can be designed to progress through any code, if output logic is included in the sequential system. There is however, one shortcoming that must be taken into account in practical circuits. The next state information may not be output from the logic simultaneously. For example a longer time is required to generate Q_2 than for Q_1 or Q_0 in the pure binary counter because of the different propagation paths. These unequal delays, together with hazards in the output logic, mean that a finite time must elapse after a pulse has been input, before the updated value of the count on the output is valid.

Hazards arise from unequal propagation paths in combinational logic. A delay of 5 nanoseconds to 1 microsecond, depending on the type of logic being used, is required for the hazards to pass.

Exercise 5.2

Design an asynchronous counter to increment in the following sequence: 000,010,011,100,111,101,001,110,000 etc.

(Investigate whether the counter can be built using next state logic only. If not, use a binary counter with output logic.)

Synchronous Sequential Logic

In a synchronous sequential logic system the switching of the circuit depends solely on the input data, but the timing within the circuit is controlled by an external clock. The clock consists of a stream of pulses, and each pulse allows the logic to switch once only, and this allows us to be more precise about the definitions of the present and next states of the system.

In an asynchronous system the present state is defined by the internal inputs **Q** at the instant external data is applied. The next state is the value of **Q** when the logic has responded to the inputs. The time taken to generate the next state depends solely on the characteristics of the logic circuits and is normally no more than a few microseconds. The next state is fed back and becomes the new present state and the circuit will operate again. In asynchronous circuits it is highly probable that the value of the internal state will be corrupted as it is fed back. This is due to unequal propagation delays in the logic.

In asynchronous circuits it is not possible to distinguish between consecutive present states or inputs which are numerically equal.

In synchronous sequential logic the present state of the system is its internal state before a clock pulse has been applied. When a clock pulse occurs, the system switches. The next state is generated and is fed back to the inputs of the logic to become the new present state. However, no further switching occurs until the next clock pulse arrives. During this time errors due to propagation delays will disappear and internal states can be precisely identified and generated at the rate of one state per clock pulse.

The period of the clock waveform must be greater than any propagation delays in the system.

Present/Next State Table

A present/next state table relates the next state of a sequential system to the external inputs and the present state. The structure of the present/next state table is given in Table 5.1.

The columns of the Table represent the value of the inputs and the rows are labelled with the present state of the system. The contents of the cells give the value

Unlike the K-map, the present/next state table can have unlimited rows and columns.

Table 5.1 A Present/Next State Table

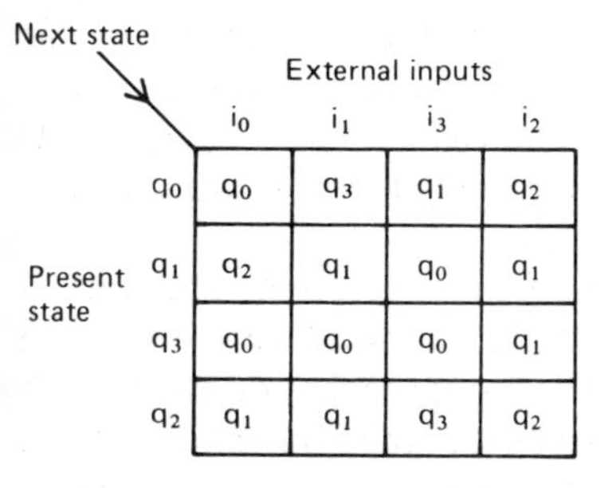

	i_0	i_1	i_3	i_2
q_0	q_0	q_3	q_1	q_2
q_1	q_2	q_1	q_0	q_1
q_3	q_0	q_0	q_0	q_1
q_2	q_1	q_1	q_3	q_2

of the next state after a clock pulse has been applied. State sequences can therefore be identified on a present/next state table.

Design of Synchronous Next State Logic

The next state logic in a synchronous sequential system can be based on either JKFFs or synchronous TFFs. The output of each device contributes 1 bit to the value of the internal state and the switching pattern for each flip-flop can be obtained from the present/next state table. Transition tables for both the JKFF and the synchronous TFF were developed in Chapter 4. By referring to the appropriate transition table for the particular type of flip-flop to be used in a system, the logic values required on the outputs of the flip-flops may be identified, and then the driving logic function obtained. The complete design process will be illustrated in the following worked example.

Worked Example 5.3 Design a timer having 3-bits resolution that measures the duration of an input pulse.

A block diagram of the system has a single external input I and three outputs to display the value of the count to 3-bits accuracy.

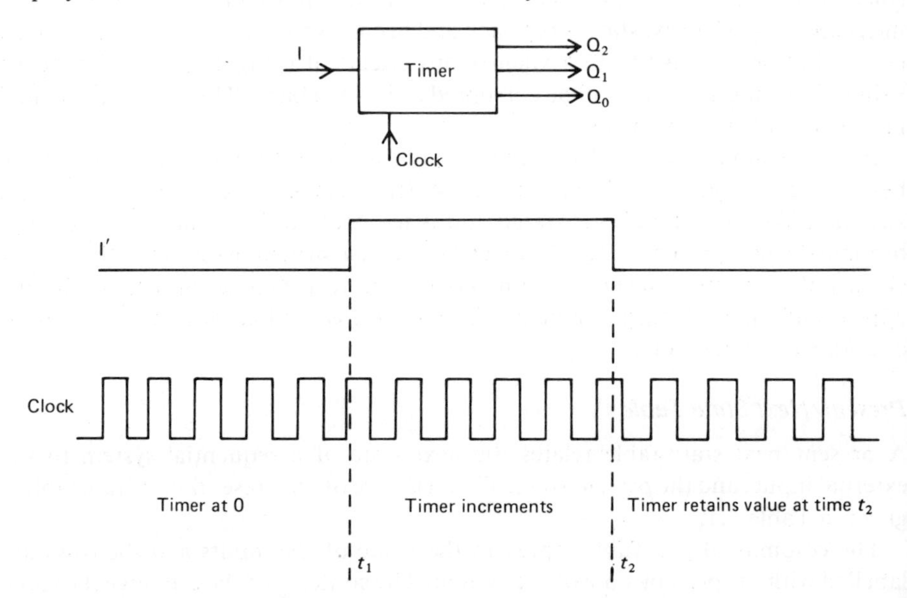

An external clock whose frequency must be known is also required.

A present/next state table for the timer can be compiled. There are two input states as the single input variable I can be either 0 or 1 and eight internal states giving the value of the count from 000 to 111. No output logic is required in this system. The outputs are identical to the internal states. The present/next state table will therefore require two columns, I = 0 and I = 1 and eight rows, $Q_2\, Q_1\, Q_0 = 000$ to 111. The contents of each cell will be the next value of $Q_2Q_1Q_0$ following a clock pulse.

Cell ($I = 0$, $Q_2Q_1Q_0 = 000$) represents the state of the timer before t_1. The clock is running but the input pulse has not yet occurred. The counter must therefore remain in state 000. When the input becomes 1 the next state of the counter at each clock pulse is the present state plus 1. For non-zero values of the count and $I = 0$, the next state of the counter needs to be the same as the present state. This condition

The unit of time in the system is the clock period.

applies after time t_2 where the input has passed and the duration of the pulse needs to be held in the system. The complete present/next state table for the timer is:

Next state

		I = 0	I = 1
Present	000	000	001
state	001	001	010
$Q_2 Q_1 Q_0$	010	010	011
	011	011	100
	100	100	101
	101	101	110
	110	110	111
	111	111	000

The designer can now establish how many flip-flops are required in the system. In general, n flip-flops can represent 2^n states and n must be an integer. Therefore given the total number of internal states x, n is the integer power of 2 equal to x or the first integer power of 2 greater than x. In this design x is 8 and $2^3 = 8$. Therefore n, the number of flip-flops required, is 3.

Either JKFFs or synchronous TFFs may be used in the next state logic. In this example we shall use synchronous triggers and the reader can repeat the design for him/herself using JKFFs. The output of each synchronous trigger contributes 1 bit to the value of the state. The present/next state table specifies the state transitions within the system for all possible values of inputs. For example, with the present state 000 and I = 0, the next state is 000. The flip-flop representing the least significant bit of the state, remains at zero. Its present to next state transition is therefore 0 to 0. However, for the same state but external input I = 1, the least significant bit must switch from 0 to 1. The output transitions of the flip-flops for all possible input conditions can therefore be obtained from the present/next state table. The inputs to the flip-flops now have to be identified to give these desired output switchings. The transition table for the synchronous TFF is:

A TFF requires one driving function (the T input). A JKFF requires two; however they are usually simpler, owing to the 'don't cares' in the JKFF transition table.

Input T	Output transition
0	$0 \rightarrow 0$
1	$0 \rightarrow 1$
0	$1 \rightarrow 1$
1	$1 \rightarrow 0$

The driving logic functions T_0 T_1 and T_2 can now be designed. They are all functions of I, Q_2, Q_1 and Q_0. A K-map may be used if the functions are required in minimal form. The map for T_2 is:

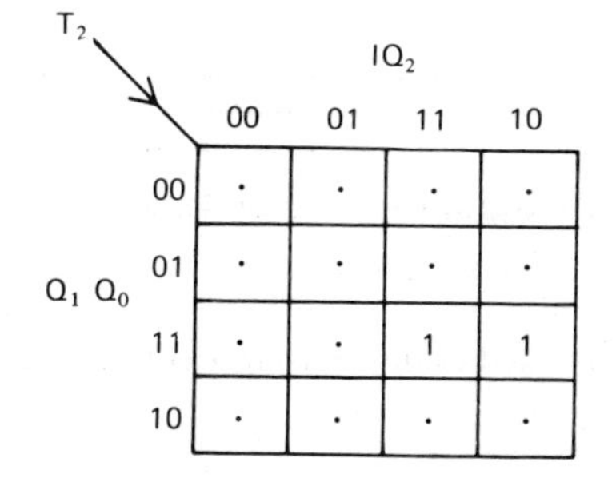

$$T_2 = I.Q_1.Q_0$$

Remember. The axes of the K-maps must be labelled in Gray code.

. on K-map indicates logical 0.

The contents of each cell of the K-map can be deduced from the system's present/next state table and flip-flop transition table. For example cell $IQ_2 = 00$, $Q_2Q_1 = 00$ on the K-map corresponds to state $I = 0$, $Q_2\,Q_1\,Q_0 = 000$ on the present next state table. The K-map represents the function T_2, which is driving the flip-flop producing the most significant bit Q_2 of the internal state. From the present/next state table we see that the present value of Q_2 is 0 and its next value is also 0. From the transition table for the TFF, 0 to 0 on the output requires T to be 0 and this is entered into the K-map. The procedure is repeated for the 15 remaining input conditions and the K-map completed. The driving function T_2 may then be obtained from the K-map. It is

$$T_2 = IQ_1\,Q_0$$

K-maps for T_1 which produces bit 1 of the internal state and T_0, which generates the least significant state bit Q_0 can also be plotted. They are:

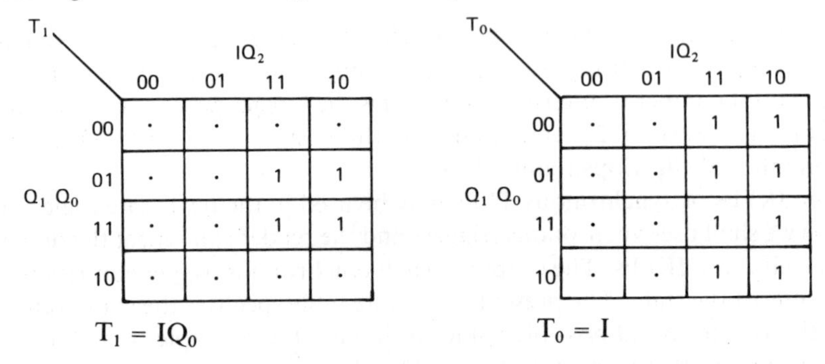

T_1: Q_1Q_0 \ IQ_2	00	01	11	10
00	·	·	·	·
01	·	·	1	1
11	·	·	1	1
10	·	·	·	·

$T_1 = IQ_0$

T_0: Q_1Q_0 \ IQ_2	00	01	11	10
00	·	·	1	1
01	·	·	1	1
11	·	·	1	1
10	·	·	1	1

$T_0 = I$

The logic diagram for the complete timer is given in Fig. 5.7.

The timer will be accurate within one clock pulse regardless of the value of the count. The accuracy will also depend on the precision of the clock frequency. In this circuit it is essential to have an external clear facility to each of the flip-flops in order to reset the timer to zero after use.

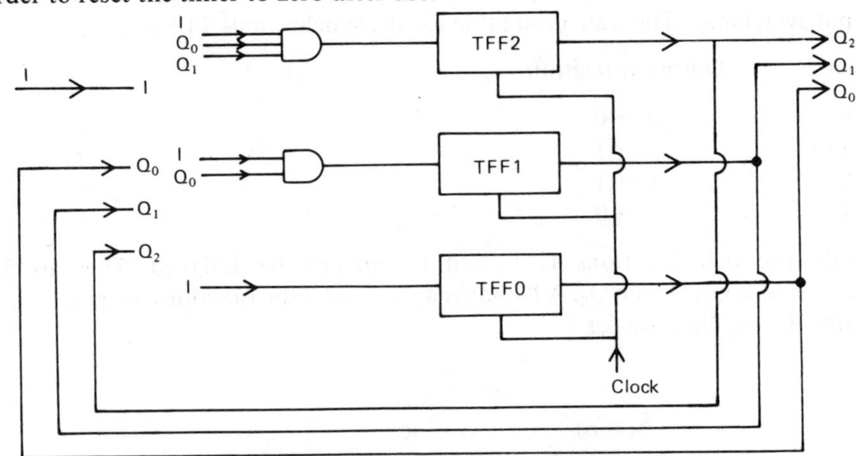

Fig. 5.7 Logic diagram for a 3-bit timer.

Exercise 5.3 Redesign the 3-bit timer in the above worked example using JKFFs in the next state logic.

Cellular Sequential Logic

In combinational logic design we saw that cellular or iterative circuit forms are very desirable as they enable the resolution in the logic system to be increased without requiring a total redesign of the system. Additional standard modules are simply added to the existing hardware. Cellular circuit forms are most useful when fabricating large systems in VLSI as the cell can be repeated automatically on the mask. The design process is reduced to the identification of a suitable cellular structure and the logical design of the cell.

A fomal design of a system having many variables would be very time-consuming.

These advantages are equally valid for cellular sequential circuits. In the timer circuit (see Worked Example 5.3) the driving logic functions were

$$T_0 = I$$
$$T_1 = IQ_0$$
$$T_2 = IQ_0Q_1$$

A regular pattern emerges from these functions. The general function T_n — the trigger function producing the *n*th significant bit of the count — is the AND of the input I and the present values of all the less significant bits:

$$T_n = IQ_{n-1}\,Q_{n-2}\ldots Q_2\,Q_1\,Q_0$$

Hence the system could be expanded to *n*-bits resolution. The *n*th flip-flop would however require a driving logic function comprising an AND gate with *n* inputs. This function would become unwieldy for large values of *n*.

Comparing T_2 with T_1 we see that

$$T_2 = IQ_0\,Q_1 = T_1\,Q_1$$
because $T_1 = IQ_0$
Therefore $T_n = Q_{n-1}\,T_{n-1}$

For any flip-flop in the timer, the driving function is the AND of the output of the previous stage Q_{n-1} and the driving function to that stage. The circuit can therefore be built out of these standard cells and the resolution depends solely on the number of cells in the system as shown in Fig. 5.8.

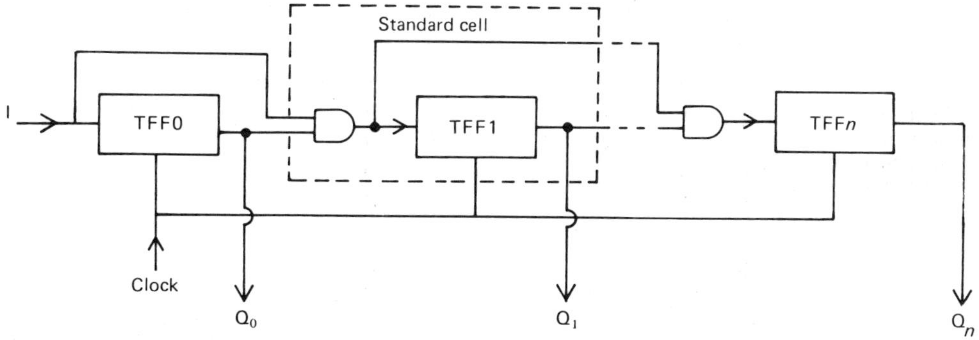

Fig. 5.8 An iterative cellular version of a timer.

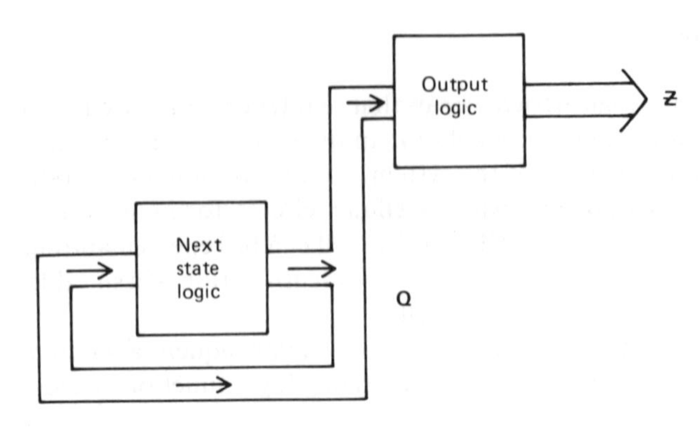

Fig. 5.9 An autonomous sequential circuit.

Autonomous Sequential Circuits

An autonomous sequential circuit does not have any external data inputs. The output logic and next state logic are functions of the internal states only. A synchronous system will, of course, require a controlling clock. The block diagram of a general autonomous sequential circuit is given in Fig. 5.9.

Code generators are a simple form of autonomous circuit and their design will be illustrated in the following worked example.

Worked Example 5.4

Design a code generator based on JKFFs, that produces the following code cycle at the rate of one word per clock pulse

000 → 001 → 011 → 110 → 100

The logic is autonomous as there is no external input to influence the next state.

Three flip-flops are required to represent the 3-bit code word. The present/next state table in an autonomous circuit consists of two columns, the present state and the next state as shown below:

Remember. d is 'don't care'. The present state does not occur in the desired sequence, so its next state is hypothetical.

Present state			Next state		
Q_2	Q_1	Q_0	Q_2	Q_1	Q_0
0	0	0	0	0	1
0	0	1	0	1	1
0	1	0	d	d	d
0	1	1	1	1	0
1	0	0	0	0	0
1	0	1	d	d	d
1	1	0	1	0	0
1	1	1	d	d	d

The 'don't care' conditions in the present/next state table are important. They occur because the code sequence contains only five different binary words. This means three flip-flops are required that are capable of representing a total of eight states. Three binary words do not occur in the code sequence. They are 010, 101 and 111. If the system functions correctly these values cannot exist as present states and therefore their corresponding next states may be regarded as 'don't cares', which can be carried through to the driving logic.

The driving functions for the flip-flops can be specified by referring to the present/next state table and the JK transition table. For flip-flop 1 (giving the least significant state bit Q_0) K-maps for driving functions J_1 and K_1 which are both the functions of Q_2 Q_1 and Q_0 are as follows:

JKFF Transition Table:

J	K	$Q_r \rightarrow Q_{r+1}$
0	d	$0 \rightarrow 0$
1	d	$0 \rightarrow 1$
d	0	$1 \rightarrow 1$
d	1	$1 \rightarrow 0$

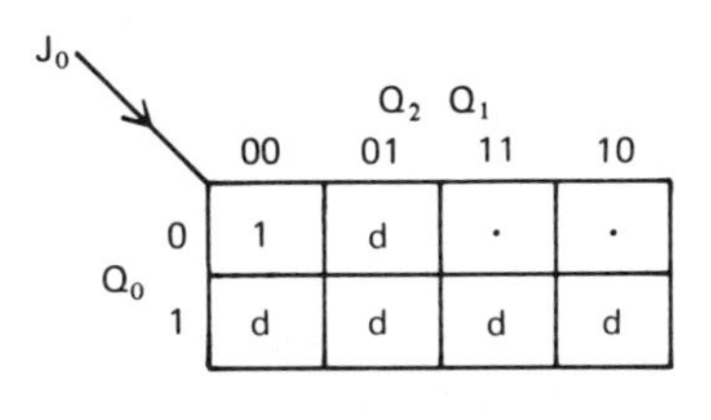

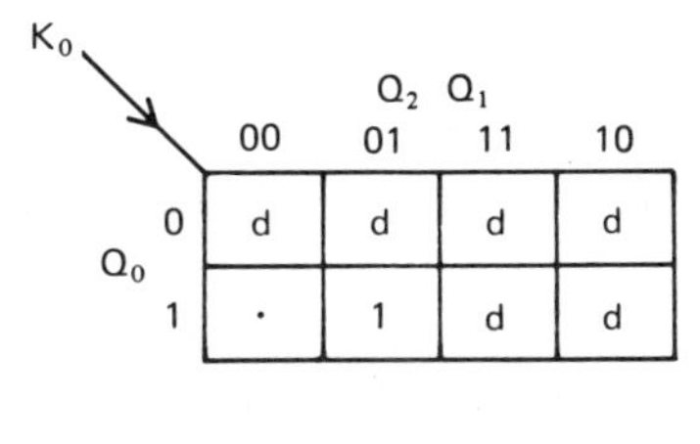

giving $J_0 = \overline{Q}_2$ $K_0 = Q_1$

By a similar process the following functions can be obtained:

$J_1 = Q_2$, $K_1 = \overline{Q}_2$
$J_2 = Q_1$, $K_2 = \overline{Q}_1$

In this example the driving logic simply consists of connections between the outputs of the flip-flops (the state variables) and their inputs. In more complicated systems the flip-flop inputs may be logical functions of their outputs.

If asynchronous data is applied to the clock, the 'code generator' type of circuit acts as a non-binary counter. This type of circuit does not have a controlling clock independent of the data. The switching is however synchronized by the input data. It is a data-synchronized circuit. It is therefore possible to use a synchronous counter, with its clock driven by input data as an alternative to asynchronous design with edge-triggered flip-flops. Referring back to Worked Example 5.1, that asynchronous system could equally well have been organized as a synchronous autonomous machine, based on JKFFs or TFFs to generate the desired count sequence. If so, the clock would be driven from the asynchronous data input I.

An example of asynchronous data is a randomly occurring sequence of pulses.

Exercise 5.4

In a simple display unit the binary words 0001 to 1111 represent the letters A to 0 respectively and 0000 represents a space.

Design a sequential system which will print out the name JOE KING continuously on the VDU.

VDU. Visual display unit.

State Transition Diagrams

In simple sequential systems one can write down the present next state table straight from the problem specification. In more complex systems where the internal states are not obvious on inspection, the system can be modelled on a state transition diagram and the input states identified. A state transition diagram is a graphical representation of a sequential system. It consists of nodes, representing the internal states, and flow lines between nodes, which are labelled with the inputs causing the state transition and the resulting output. The state transition diagram for the JKFF is given in Fig. 5.10. There are two nodes q_0 and q_1 representing the internal states 0

This form of state transition diagram is known as the Mealy model. Another standard model is due to Moore, and has its nodes labelled with the internal state and the resulting output.

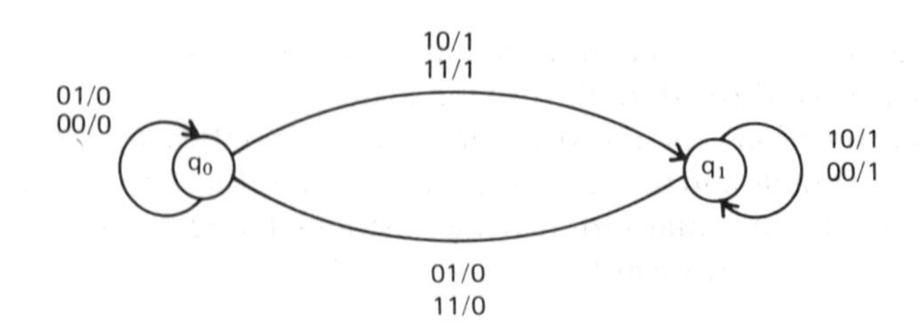

Fig. 5.10 A state transition diagram for a JKFF.

and 1. The flow lines are labelled JK/Q where J and K are the inputs and Q is the output.

If the flip-flop is in state 0 (q_0) and J = 0 and K = 1, the next state of the system is 0. This is represented by the flow line leaving q_0 and returning to q_0. If however, J = 1 and K = 0, the flip-flop will be set to 1 and its output goes to 1. This condition is represented by the flow line leaving q_0 directed towards q_1.

The present/next state table can be obtained from the state transition diagram. Values of the inputs are given on the flow line labels.

A flow line may connect different nodes or form a loop leaving and then re-entering the same node.

Every flow line starts from, and terminates in, a node. The starting node represents the present state and the final node the next state. In Fig. 5.10, if q_0 is the present state and the input is 10 the next state will be q_1 whereas present state q_1 and input 10 results in next state q_1.

The present/next state table can be produced by considering all the transitions between nodes on the state transition diagram. For the sequential system in Fig. 5.10 the table is:

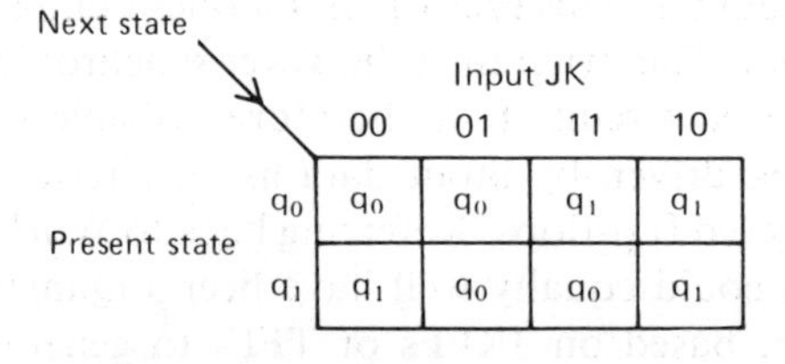

Present state	00	01	11	10
q_0	q_0	q_0	q_1	q_1
q_1	q_1	q_0	q_0	q_1

The assignment of binary values to the states is arbitrary, but some state assignments may lead to marginally simpler logic circuits than others.

At this stage the states have to be assigned binary values. The next state logic can then be designed using the present/next state table. As we started with a JKFF, the reader should find that the next state logic consists of one flip-flop and the inputs are identical to the variables labelled on the state transition diagram, where q_0 is 0 and q_1 is 1.

Worked Example 5.5

Design a sequential logic circuit that will detect the sequence 101 in a stream of serial data, input at the rate of 1 bit per clock pulse.

The circuit has a single input variable I which is either 0 or 1, and a single output Z that becomes 1 when the input sequence 101 has been received. However, it is not obvious at the outset how many internal states are required. The internal states are identifiable from a state transition diagram.

Let internal state A be the initial state where none of the input bits have arrived in the correct sequence.

In state A, the external input can be either 0 or 1. If I = 0 we remain in state A as

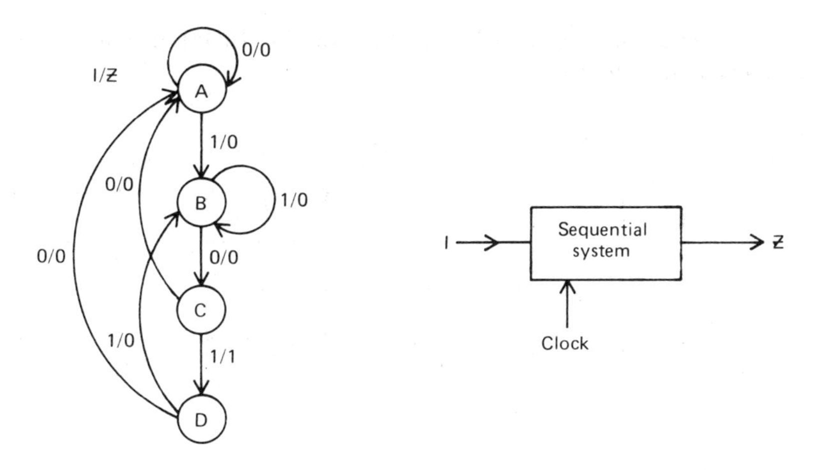

Fig. 5.11 State transition diagram of the sequence detector.

the input does not correspond with the first bit in the sequence we wish to detect. If, however, I = 1 the first bit in the sequence has occurred. This condition is represented by new internal state B. The output in both cases is zero as the complete sequence has not yet been detected. When in state B further input data can be received. If I = 0 the second bit of the sequence 101 has arrived and we switch to state C. If I = 1 the current input might still be the first bit of the sequence 101 and the system should remain in B. The complete state transition diagram is given in Fig. 5.11.

The state diagram contains four nodes and state D is entered after the complete input sequence has been detected. The present/next state table is:

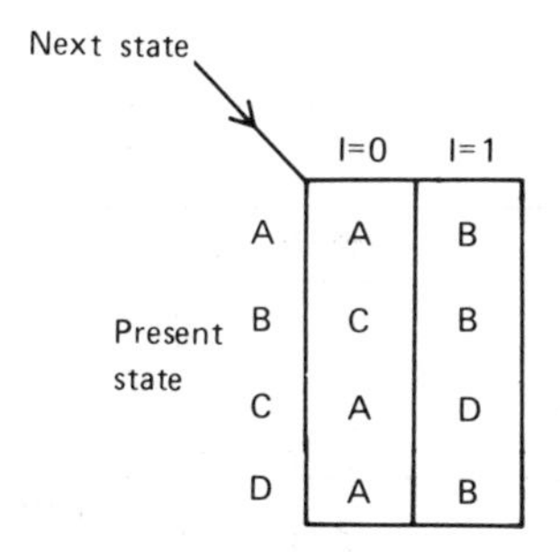

Present state	Next state I=0	Next state I=1
A	A	B
B	C	B
C	A	D
D	A	B

Assigning binary values to the internal states (let A = 00, B = 01, C = 11 and D = 10) we obtain:

As the assignment of binary values is arbitrary, there are a large number of different hardware solutions to this problem. The state is made up from the outputs of two flip-flops Q_1Q_0.

Present $Q_1\ Q_0$	Next I = 0	Next I = 1
00	00	01
01	11	01
11	00	10
10	00	01

2^n states require n flip-flops.

The next state logic requires 2 JKFFs and the driving functions may be derived from K-maps as shown below:

Refer to Worked Example 5.4 for derivation of K-maps from state tables.

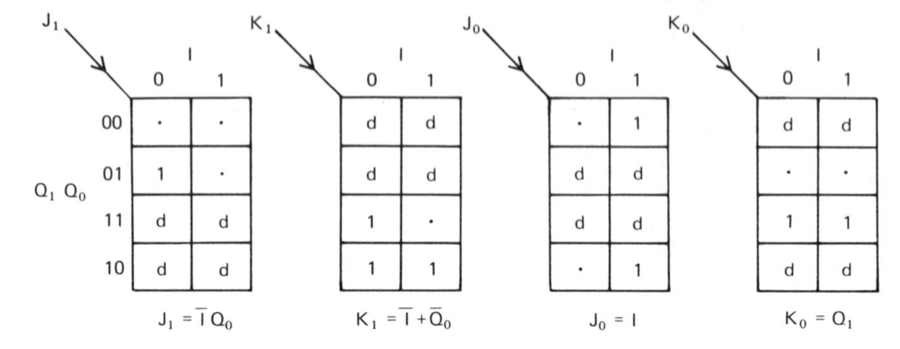

See Fig. 5.1.

In this system output logic is required as the output is not identical to the value of the internal states. From the general form of the sequential circuit we see that the output is a function of both the input and the internal state variables.

It is evident from the state transition diagram that the output becomes 1 when in present state C and the input is 1 (the flow line from C to D).

The output function is therefore

An alternative output function is $Z = Q_1\bar{Q}_0$.

$$Z = IQ_1Q_0$$

The logic diagram for the complete system is given in Fig. 5.12.

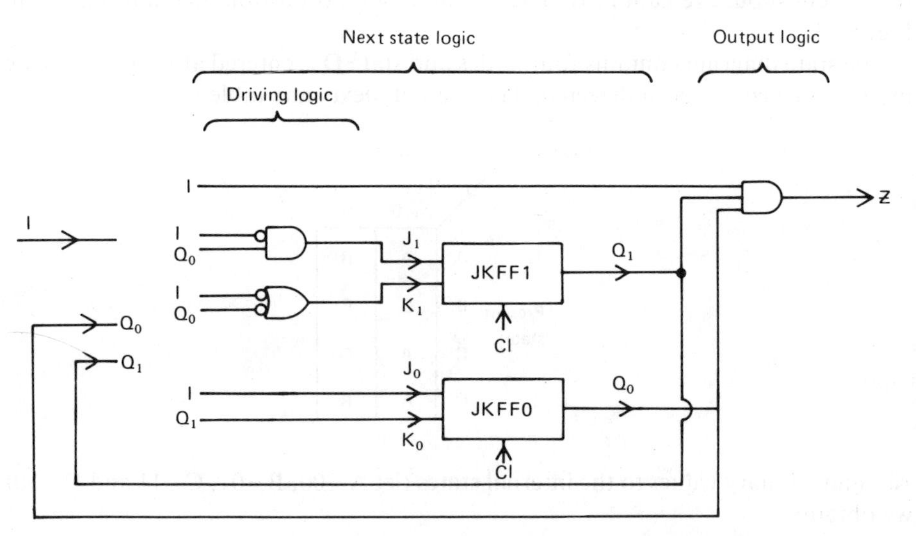

Fig. 5.12 Circuit diagram of the sequence detector.

Exercise 5.5 Design a sequential system which detects the bit-pattern 1101 in a stream of bit-serial data.

State Minimization in Sequential Logic

When modelling sequential systems on a state transition diagram it is important to have as few internal states as possible. This leads to simpler and cheaper hardware.

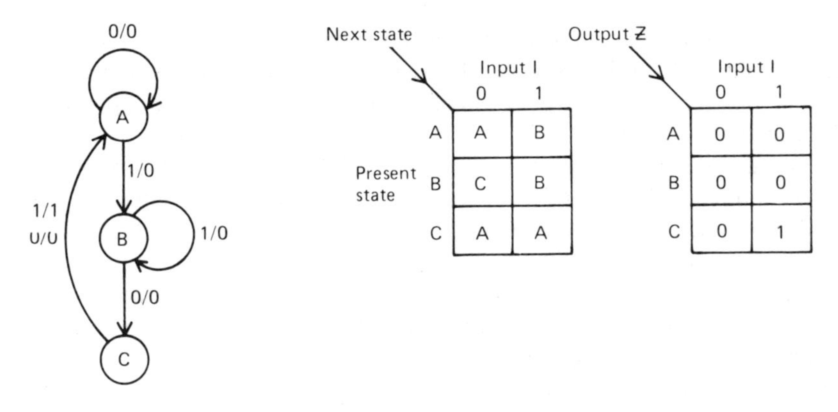

Fig. 5.13 A minimized state transition diagram and tables for the 101 sequence detector.

A state transition diagram can be minimized by combining two or more states provided their next states and outputs are identical. In the 101 sequence detector (Worked Example 5.5) comparing states A and D we see the next state is always A if I is 0, and B, if I is 1, and the output is always 0. States A and D therefore have identical next states and outputs and so may be combined into a single state. State D is then eliminated from the system and any flow line directed to it can be re-routed to A. The state minimized system is shown in Fig. 5.13.

In this minimized system only three states are required. Two JKFFs are still necessary but the driving logic is simpler as the fourth state which could be represented but is not now needed, becomes a 'don't care'.

Exercise 5.6. Design a sequential system, using 2 JKFFs and driving logic, that detects the pattern 1011 in a stream of bit-serial data.

Summary

A general sequential circuit contains two distinct logic functions: the next state logic and the output logic. Three sets of data can be identified: **I** represents the inputs, **Q** the internal states, and **Z** the outputs.

In many simple sequential systems the outputs are identical to the internal states and no output logic is required. Sequential logic can be either synchronous or asynchronous. In asynchronous systems, timing and propagation delays can cause malfunctioning, although the circuits will operate at high speed. When designing sequential logic using edge-triggered flips-flops, suitable falling edges must be identified within the system to give the desired switching sequences. In some circumstances these edges may not exist or cannot be identified and output logic is then required to encode the state sequence of a counter which can be built.

The basic building block for discrete synchronous logic is the JKFF. The present/next state table provides a suitable means of specifying the state sequences within a system. The driving logic for the JKFF can then be obtained from the JK transition table and the system's present/next state table. The inputs to the driving functions and the output logic are the state variables and the external inputs.

An autonomous sequential system has no external inputs and if the controlling clock is replaced by a data source, the JKFF can operate on asynchronous data. There is no separate clock in these systems and the logic operations are synchronized by the inputs.

The state transition diagram is a powerful tool, enabling a system to be represented graphically. This allows the properties of a sequential system to be identified and scrutinized. The present/next state table can then be deduced from the transition diagram.

Sequential systems can be state-minimized. If two or more states lead to identical next states and have consistent outputs they may be combined into a single state. The complexity of the next state logic can therefore be reduced.

For large complex systems, computer-aided design may be necessary. In practice, owing to the cost of hardware being a small item in any design proramme, the minimization of sequential logic may not be cost-effective. In VLSI design however, it could be advantageous to minimize the logic. If the basic cell of an iterative system is designed in its simplest form the total number of cells that can be fabricated on a given area of silicon is increased. Logic minimization may therefore lead to more powerful single chip systems.

Problems

5.1 Design a binary counter using asynchronous TFFs and driving logic which will count in 8421 BCD, a series of randomly occurring pulses on its input I.

5.2 Design a 4-bit walking code generator, using a 3-bit counter and output logic. (Walking code is 0000, 0001, 0011, 0111, 1111, 1110, 1100, 1000, 0000, etc.). Redesign the system, using shift registers and compare the circuits.

5.3 Using JKFFs, design a system that will output the following 2-bit code: 00, 01, 11, 10 and then continuously repeat the sequence from 01. Repeat the design using (i) TFFs and (ii) DFFs.

5.4 Design a logic system to operate a set of traffic lights. The lights must be switched on in the following sequence: red, red and amber, green, amber, red, etc., and a logical 1 from the controller switches a light on. The system is controlled by the following waveform:

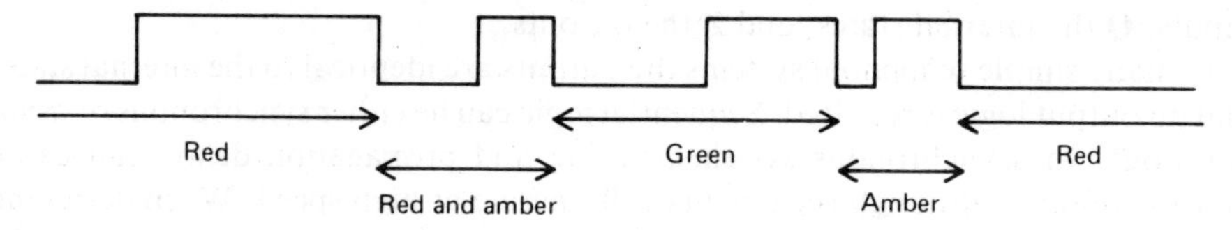

5.5 A faulty batch of JKFFs have the following state transition diagram:

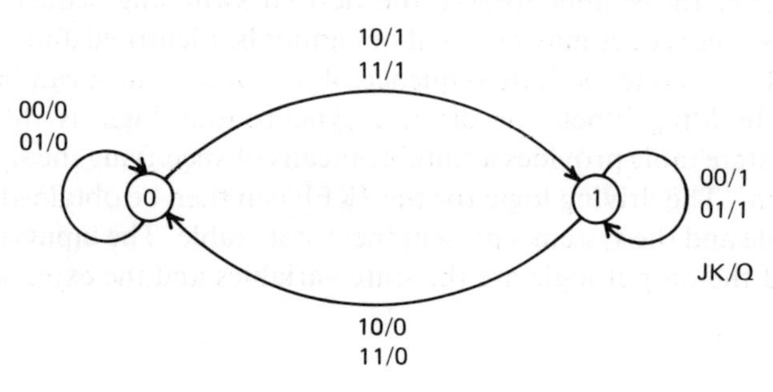

Obtain the transition table for the faulty flip-flops. How should the input terminals be relabelled so that the devices could be sold as standard flip-flops? What type would they be?

5.6 Design a logic system to implement the following state transition diagram:

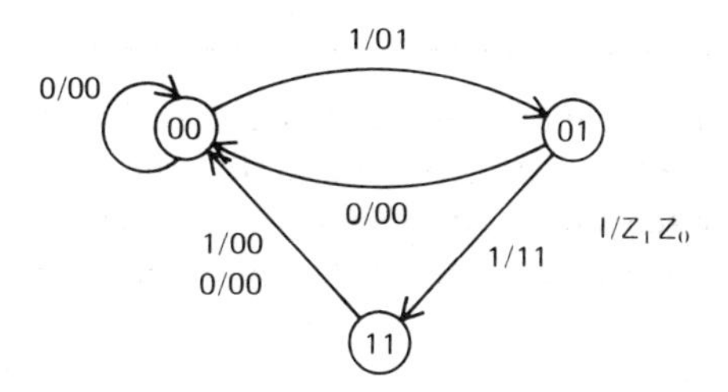

Describe in words, the behaviour of the circuit.

5.7 State minimize the following sequential machine:

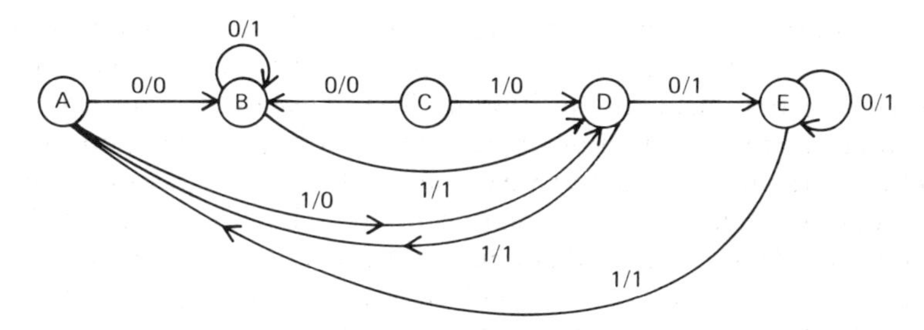

5.8 Design a synchronous sequential logic system, using the minimum number of JKFFs together with NAND gates for driving and output logic, that will provide an output $Z=1$, as long as the repetitive sequence AB = 00, 01, 11, 10, 00, etc., is applied to the inputs. (One 2-bit word is input on each clock pulse). If any input is applied out of sequence, the output must go to 0. The output should become 1 again only if the next correct input is applied.

5.9 Design an entry system that provides a signal to unlock a door provided the input sequence AB = 00, 10, 01, 11 is applied and sounds an alarm continuously if any deviation occurs on the inputs. Comment on any possible timing problems.

5.10 Design a machine that compares pairs of 4-bit numbers input bit-serially on two lines A and B, and outputs a signal whenever two words are identical.

6 The Digital System

Objectives

- ☐ To examine the influence of integrated circuit technology on logic design.
- ☐ To investigate the concept of programmable logic.
- ☐ To design logic systems based on memories and arrays.
- ☐ To specify a system as a set of registers and its operation as a series of register transfers.
- ☐ To examine the bus concept.
- ☐ To introduce the basic principles of fault diagnosis in a digital circuit.

The development of integrated circuit technology has had a profound influence on logic design. In the 1950s all logic gates were built with discrete components — transistors, resistors, capacitors and the like. The advent of the transistor, a decade before, had made the construction of individual gates a practical proposition. The late 1960s saw the introduction of very small scale integrated circuits containing less than one hundred electronic components per chip. A typical product that appeared in the 1960s is the quad 2-input NAND package — a 14 pin dual in-line integrated circuit that contains four NAND gates, each having two inputs. These circuits enable logic systems to be constructed from discrete gates.

Since the introduction of integrated circuits in the 1960s, dramatic developments have taken place. The complexity has, on average, doubled every three years and we now have very large scale devices that contain the equivalent of 100 000 components. The cost per elementary logic function however has halved, in real terms, every year, giving a cost reduction of one million times in 20 years. This can best be illustrated by considering memory. In 1965, 1 bit of semiconductor memory comprised a DFF made out of discrete components. The cost would have been the equivalent of a graduate engineer's pay for 20 hours work. Today, 1 bit of memory is one quarter millionth of a 256K-bit memory chip and the cost of the whole chip is less than $100. The bit cost is equivalent to an engineer's pay for a fraction of a second.

A wide range of dedicated logic integrated circuits are now available. The reader is directed to the manufacturers' catalogues for further details. In this chapter we will be examining integrated circuits that are not restricted to single function operations and that can be used in the design of a general digital system.

Programmable Logic

A dedicated logic circuit will only perform the function it was designed to implement. A combinational circuit behaves according to its truth table, whereas a sequential circuit's characteristics are governed by its state table. In a programmable logic system, a standard hardware structure is electronically reconfigurable so that the different logic operations can be performed, without having to change the circuit physically.

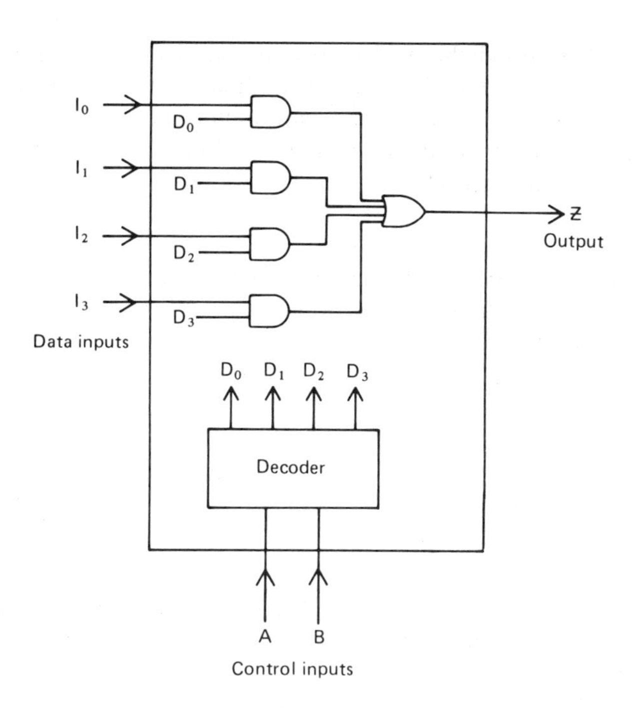

Fig. 6.1 Logic diagram of a 4 to 1 multiplexer.

The Multiplexer

The multiplexer is a multi-input, single-output circuit whose prime function is to convert parallel data to serial data. It differs from the parallel input/serial output shift register in one important respect — it has additional control inputs that address individual data input lines. The inputs can therefore be output in any order dependent on the control address.

Figure 6.1 shows a 4 to 1 multiplexer. The control inputs must be able to address any input line, hence two control inputs are required by this multiplexer.

The control inputs enter a decoder that has four outputs. Each responds to a specific input value on AB giving:

$$D_0 = \overline{A}\,\overline{B}$$
$$D_1 = \overline{A}\,B$$
$$D_2 = A\,\overline{B}$$
and $D_3 = A\,B$

Hence for a given control input, only one decoder output will be at logical 1. The decoder outputs are ANDed with corresponding data inputs I_i and then ORed together to give the multiplexer output Z. Thus the output Z takes on the value of I_i, where *i* is the address AB, applied to the control inputs. Hence the multiplexer is a parallel to serial data converter.

If the addresses are driven from a pure binary counter, the multiplexer acts as a shift register. It does not, however, have any internal storage; therefore the inputs must be held constant during operation.

The multiplexer can also be used as a universal combinational logic function. Any logic function of n variables can be set up in a 2^n to 1 multiplexer. The control inputs become the variables of the logic function and the data inputs are driven from logical switches. The truth table of the function to be programmed into the multiplexer must be evaluated and each minterm which requires an output set to 1, identified. The logical switches which are addressed by the minterms, are set to 1 and all other data inputs remain at 0. The multiplexer will then perform the combinational function as specified by the truth table. The same multiplexer can be programmed to perform another combinational function simply by changing the setting of the data input switches.

Worked Example 6.1

Implement an EXCLUSIVE-OR function in a 4 to 1 multiplexer. Reprogram the multiplexer to perform the AND function.

Truth table for EX.OR is

AB	F
00	0
01	1
10	1
11	0

Minterms of EX.OR are $A\overline{B}$ and $\overline{A}B$. If A and B are applied to the control inputs of the multiplexer, data inputs I_1 and I_2 will be selected by the minterms. These inputs must be set to 1.

Hence the multiplexer circuit needed to implement EX.OR is:

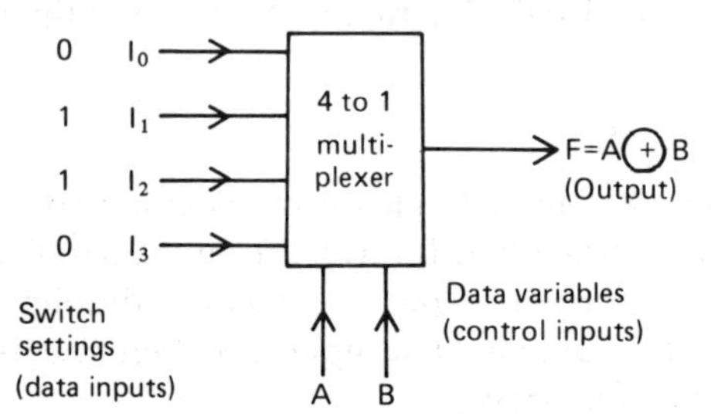

For the AND function, we only have one minterm (AB). Therefore data input I_3 must be set to 1 giving

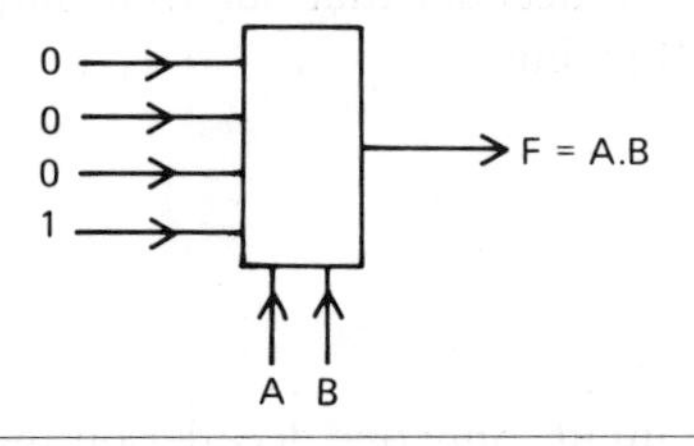

A 2^n to 1 multiplexer can be programmed to perform any function of n variables. This means that a 4 to 1 multiplexer can implement 16 functions of 2 variables, and with a 16 to 1 multiplexer we can perform any one to the 65 536 possible functions of 4 variables. There is, however, a physical limitation of the number of data inputs that are available on an integrated circuit, and this will limit the complexity of the combinational functions that may be programmed into the multiplexer. It is possible to overcome this restriction by allowing either logical variables or logic functions to be input to the data lines. The latter can be developed as a multi-level multiplexer system.

The 16 functions are the binary connectives (see Chapter 2).

↑ n inputs, 2^n input states, 2^{2^n} different output sets.

A 10-variabled function would require a single multiplexer with 1024 (2^{10}) data inputs. This is physically impossible.

Consider the 3-variabled function:

$$F(ABC) = \Sigma(1,2,4,6) \tag{6.1}$$

which we wish to implement in a 4 to 1 multiplexer. This can be achieved by making one variable available (say A) to the data inputs, and B and C to the control inputs.

The full Boolean equation of the function is

$$F = \overline{A}\,\overline{B}\,C + \overline{A}\,B\,\overline{C} + A\,\overline{B}\,\overline{C} + A\,B\,\overline{C} \tag{6.2}$$

Equation 6.2 must be rearranged so that it contains all possible terms of BC, the variables being applied to the control inputs. Thus

Terms are $\overline{B}\,\overline{C}$, $\overline{B}C$, $B\overline{C}$ and BC.

$$F = (A)\,\overline{B}\,\overline{C} + (\overline{A})\,\overline{B}\,C + (A+\overline{A})B\overline{C} + (0)BC$$
$$F = (A)\overline{B}\,\overline{C} + (\overline{A})\overline{B}\,C + (1)\,B\overline{C} + (0)\,BC \tag{6.3}$$

$(A + \overline{A}) = 1$.

The bracketed terms are called residues and can be either logical constants or the variable A or the inverse $\overline{A}$. Variables B and C are applied to the control inputs of the multiplexer and the residues to the data inputs. The multiplexer circuit to implement Equation 6.1 is therefore:

If the function does not have minterms containing all combinations of B and C (e.g., BC), then the corresponding residue is 0, ((0)BC), giving a 0 output.

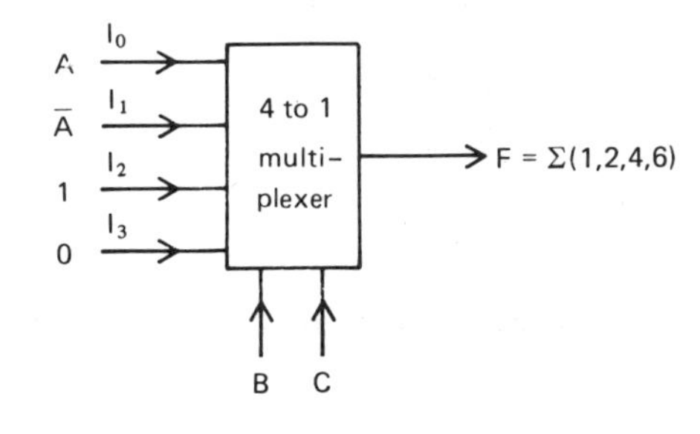

and a 4 to 1 multiplexer implements a 3-variabled function.

If the data inputs are restricted to 0 and 1, the 4 to 1 multiplexer can only act as a 2-variabled function.

Exercise 6.1

Program an 8 to 1 multiplexer to perform the function

$$F = \Sigma(1,2,4,7,8,9,14,15)$$

Multilayer Multiplexer Systems

In a multilayer multiplexer system, logic functions are generated by multiplexers, which are then input to a second and subsequent layer of multiplexers. The size of the multiplexers and the number of layers will depend on the complexity of the function to be performed. In the following analysis we will confine ourselves to a 2-layer system of 4 to 1 multiplexers which can be programmed for 5-variabled functions (Fig. 6.2). The methodology may be extended for larger systems.

The function F(VWXYZ) is first expanded universally about two variables (say Y and Z). The residues provide the data inputs to M_1. Each residue is then expanded about two variables (say W and X) and its residues, which form the inputs to the first layer, determined. The multiplexers in the first layer all have common control inputs. Hence the function can be set up by programming the 1st layer data inputs with the appropriate settings (0, 1, V or $\overline{V}$) and applying the other four function variables to the control inputs of the multiplexers.

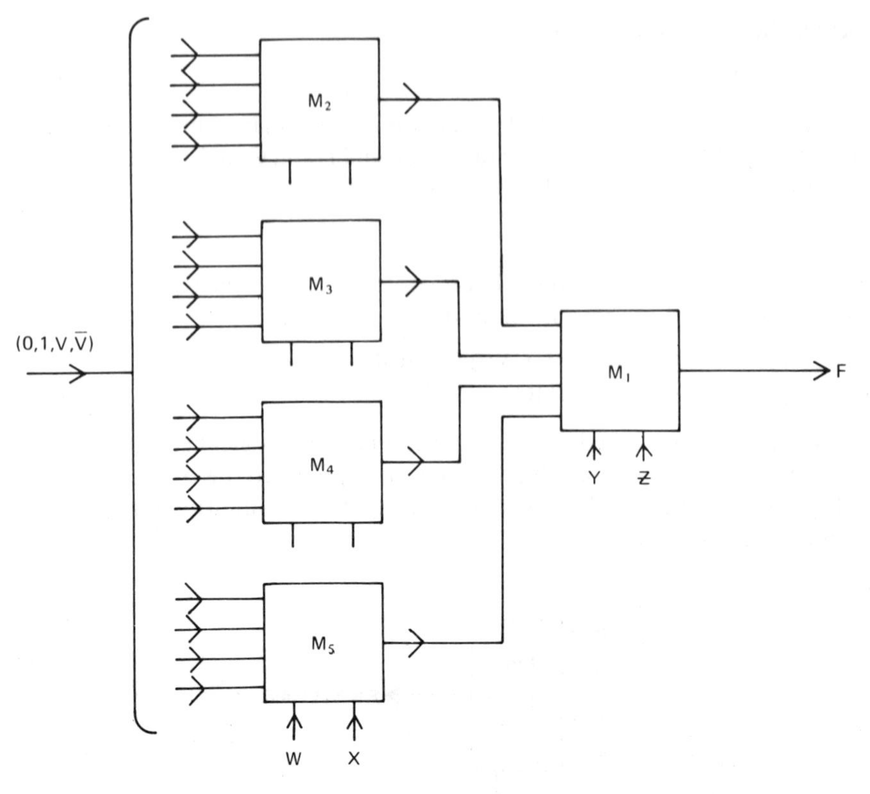

Fig. 6.2 A 2-layer system of 4 to 1 multiplexers.

Worked Example 6.2 Implement the 5-variabled function

$$F = \Sigma(1,2,4,5,7,8,12,17,19,21,28)$$

in a 2-layer system of 4 to 1 multiplexers.

The equation of the function is

$$\begin{aligned} F = {} & \overline{V}\,\overline{W}\,\overline{X}\,\overline{Y}Z + \overline{V}\,\overline{W}\,\overline{X}Y\overline{Z} + \overline{V}\,\overline{W}X\overline{Y}\,\overline{Z} + \overline{V}\,\overline{W}X\overline{Y}Z + \overline{V}\,\overline{W}XYZ + \\ & \overline{V}W\overline{X}\,\overline{Y}\,\overline{Z} + \overline{V}WX\overline{Y}\,\overline{Z} + V\overline{W}\,\overline{X}\,\overline{Y}Z + V\overline{W}\,\overline{X}YZ + \\ & V\overline{W}X\overline{Y}Z + VWX\overline{Y}\,\overline{Z} \end{aligned}$$

Expanding F about YZ gives

$$F = (F_1)\overline{Y}\,\overline{Z} + (F_2)\overline{Y}Z + (F_3)Y\overline{Z} + (F_4)YZ$$

where

$$\begin{aligned} F_1 &= (\overline{V}\,\overline{W}X + \overline{V}W\overline{X} + \overline{V}WX + VWX) \\ F_2 &= (\overline{V}\,\overline{W}\,\overline{X} + \overline{V}\,\overline{W}X + V\overline{W}\,\overline{X} + V\overline{W}X) \\ F_3 &= (\overline{V}\,\overline{W}\,\overline{X}) \\ F_4 &= (\overline{V}\,\overline{W}X + V\overline{W}\,\overline{X}) \end{aligned}$$

Expanding F_1, F_2, F_3 and F_4 about WX gives

$$\begin{aligned} F_1 &= (0)\overline{W}\,\overline{X} + (\overline{V})\overline{W}X + (\overline{V})W\overline{X} + (1)WX \\ F_2 &= (1)\overline{W}\,\overline{X} + (1)\overline{W}X + (0)W\overline{X} + (0)WX \\ F_3 &= (\overline{V})\overline{W}\,\overline{X} + (0)\overline{W}X + (0)W\overline{X} + (0)WX \\ F_4 &= (V)\overline{W}\,\overline{X} + (\overline{V})\overline{W}X + (0)W\overline{X} + (0)WX \end{aligned}$$

The multilayer multiplexer net will therefore be

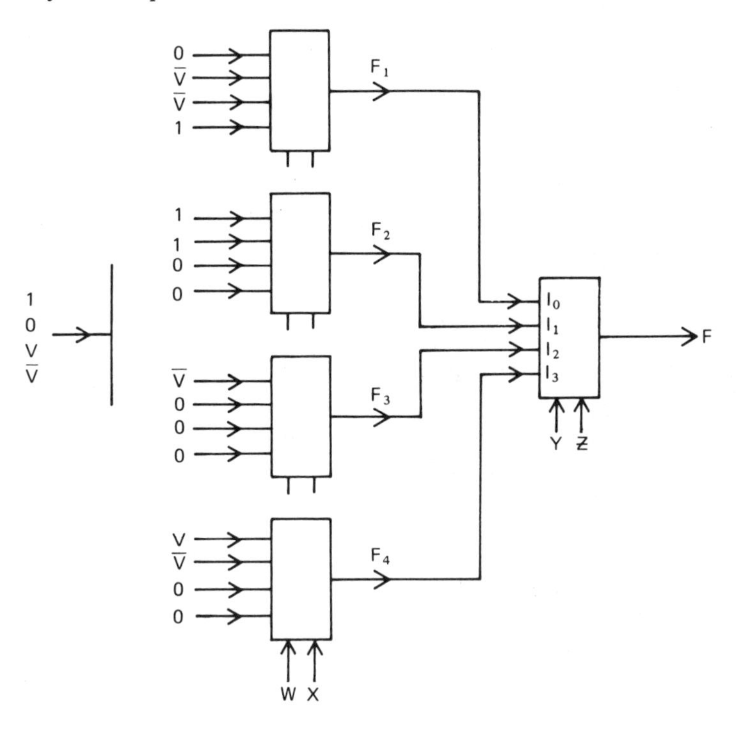

The Memory Element

The memory element is a dual-purpose device. Its primary role is that of an electronic store for binary data, and as such was discussed in Chapter 2. Memories, however, are playing an increasingly important role as programmable logic devices.

A memory circuit contains several storage elements that can be addressed by a decoder circuit. The functional diagram of a 4-bit random access memory is shown in Fig. 6.3. The decoder is functionally identical to the multiplexer decoder and the storage elements are DFFs. To write into the memory the data is applied to the 'data-in' terminal which is connected to all the inputs of every DFF. The address at which the data is to be stored is supplied to inputs A and B. This will set one, and only one, of the decoder outputs to 1. Suppose we wish to write 1 into location 0. Address AB = 00 must be applied to the address inputs and decoder outputs D_0 will be set to 1. The read/write control must also be set to 1 and the inputs to AND gate 1 will both be high, giving a 1 on the output which drives the clock of DFF 0, causing the logical value on 'data-in' to be entered into the flip-flop DFF 0. All the remaining flip-flop decoders will be at 0 and cannot therefore store 'data-in'.

To read from the memory, the read/write control is set to 0 and inverter 9 will transmit a logical 1 to all the output AND gates (5 to 8) each of which also has as inputs, a DFF output and its decoded address. The AND gate outputs are ORed together to produce the memory output whose value is the contents of the flip-flop selected by the address decoder.

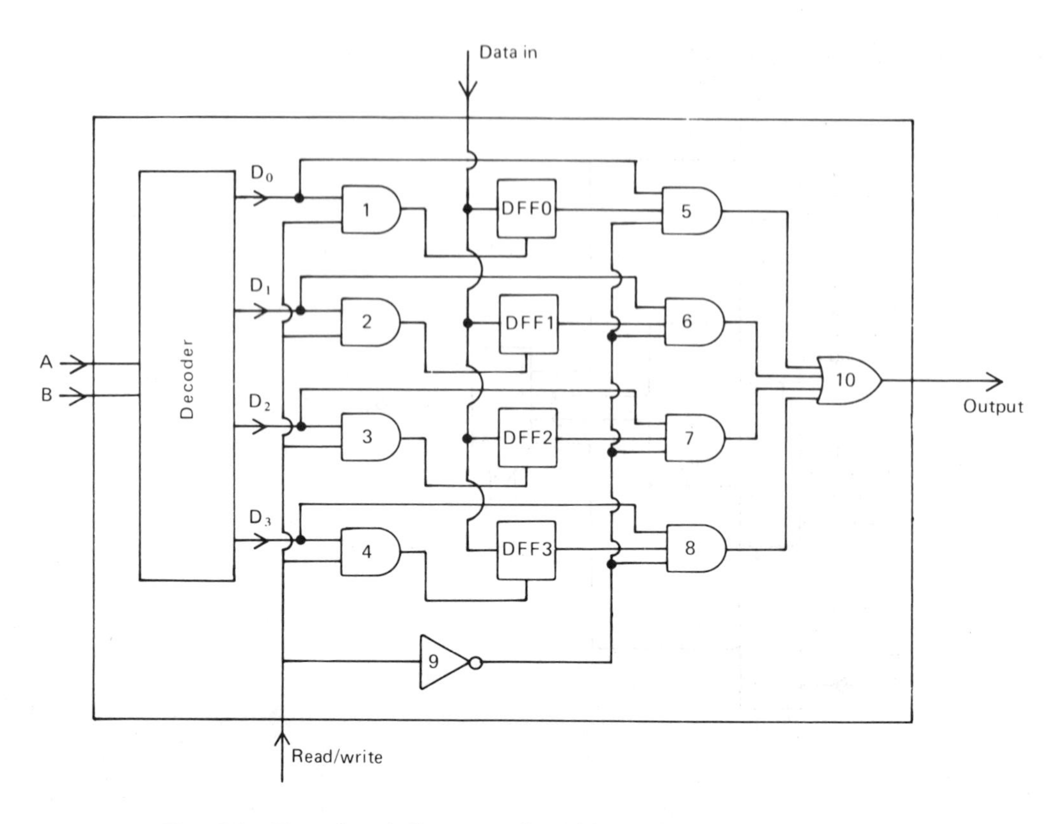

Fig. 6.3 Functional diagram of a 4-bit random access memory.

Known as 'RAM'

With this logical structure, any storage location can be accessed, depending on the value of the address. The memory locations may be read and written to in any sequence without having to be scanned, hence the name 'random access memory'.

Memory chips are available in sizes up to 256k-bits which require 18 address lines. They can also be byte-organized, where a given address will access more than 1 bit in parallel.

An *n*-bit byte organized memory uses n memory planes operating off a common decoder, as shown in Fig. 6.4. The data input and output channels are now *n* bits wide. Commercially available memories are discussed further in Chapter 7.

A memory can act as a logic function simply by storing a truth table, and as any truth table can be loaded into store, the device is both programmable and universal. A bit-organized memory only has one data input line, irrespective of its size, through which each and every storage location can be entered. This contrasts with the multiplexer which requires 2^n data inputs for n address lines. A memory is therefore capable of implementing large logic systems and is an attractive alternative to multilayer multiplexer systems.

The memory is initially cleared — all locations are reset to 0.

A memory is programmed to perform a combinational logic function by first obtaining the complete truth table of the function. The logic variables will determine the number of address lines and hence the size of the memory required. Each address line represents one input variable. For each input state in the truth table that requires an output at 1, the inputs are applied to the address lines and the location accessed is set to 1. The memory, in read mode, will then perform the logic

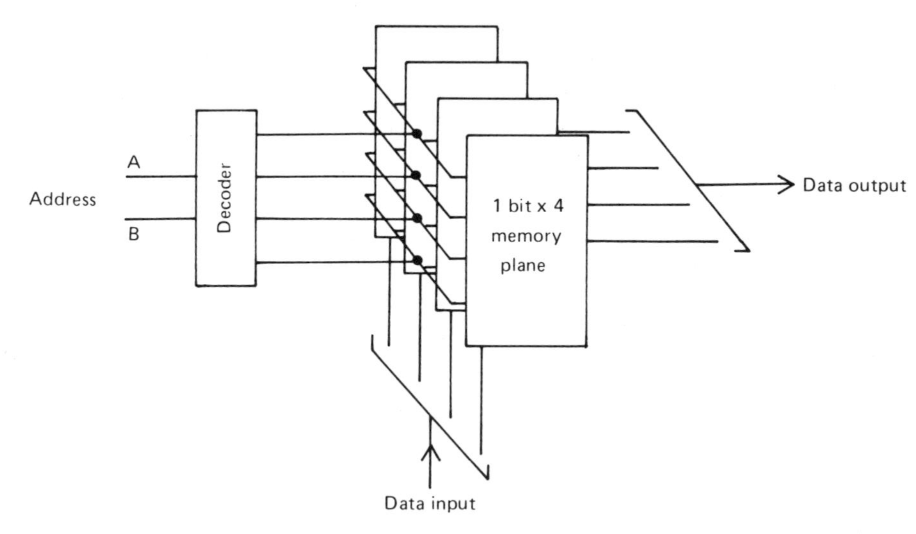

Fig. 6.4 A4 × 4-bit byte random access memory.

function specified by the truth table. Each input state will address a memory location whose contents correspond with the output condition on the truth table.

When using memories as logic functions, there is no real need to, or advantage in, minimizing the logic function beyond the elimination of any redundant variables. The size of the memory is determined by the number of dependent variables in the function, and any minimization that does not result in the total elimination of one or more variables from the function, will not reduce the storage requirements.

$F = ABC + AB\bar{C}$ requires an 8-bit memory, but if simplified to $F = AB$, a 4-bit memory will suffice.

An n-variabled function requires 2^n bit memory.

Worked Example 6.3

Design a combinational logic system based on a memory, which will indicate when a 3-bit binary number $\mathbf{X}$ is in the range $3_{10} \leqslant \mathbf{X} \leqslant 5_{10}$.

The truth table for the range detector $F = f(X_2\,X_1\,X_0)$ where $\mathbf{X} = X_2\,X_1\,X_0$ is:

X_2	X_1	X_0	F
0	0	0	0
0	0	1	0
0	1	0	0
0	1	1	1
1	0	0	1
1	0	1	1
1	1	0	0
1	1	1	0

F is dependent on all 3 input variables $X_2\,X_1$ and X_0; therefore an 8-bit memory is needed. The truth table may be loaded, by first clearing the memory and then writing logical 1 in addresses 011, 100 and 101. The address lines are driven by the input variables and when in the read mode, the memory output gives the desired functions. The memory configuration is

This may be checked on a K-map.

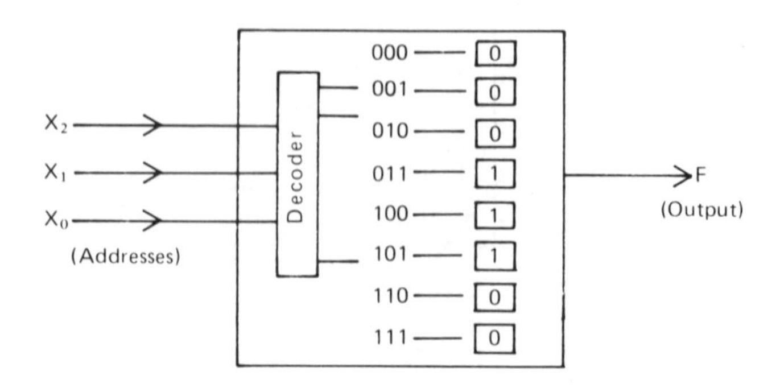

Byte-organized memories may be set to implement multi-output functions. For example, in a BCD to Gray code converter each input in BCD would address a 4-bit location where the equivalent Gray value is stored.

Furthermore, the address field may be partitioned into two parts, to enable more than one function to be programmed into a large memory. Consider a typical commercial memory of 1k-bits. It will have 10 address lines and can be used as a 10-variabled function. Suppose however, we have a number of 4-variabled functions to perform. The address field can be divided into two. Four lines are required for the input data variables, leaving 6 lines to be used as the function addresses, as shown in Fig. 6.5.

The same data input lines are used for every function. The memory address is made up of the data inputs and the function address.

The first function $F_0 = f(ABCD)$ is loaded into memory by setting the function address to 000 000 and entering the minterms via ABCD. A second function $F_1 = f(ABCD)$ can be loaded into a different section of memory by setting the function address to 000 001. The minterms of every function are entered via ABCD. Thus a total of 2^6 or 64 different 4-variabled functions can be loaded into the memory.

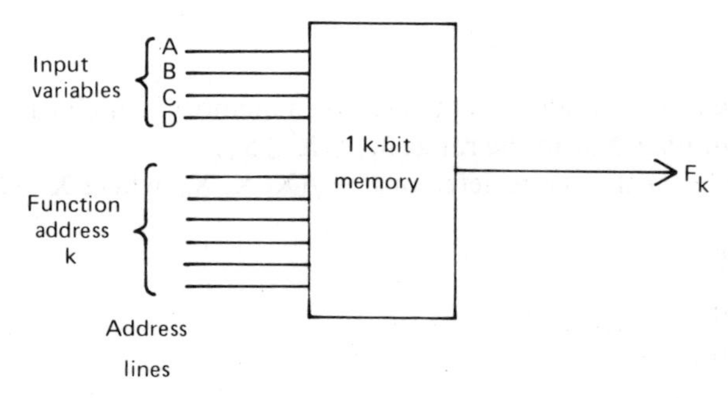

Fig. 6.5 A 1k-bit memory partitioned to perform 64 functions of four inputs.

In use, the desired function is selected via the function address. The input data is applied to ABCD and the memory output gives the function required. Different functions may be selected by changing the function address, although it must be held constant when any given function is being accessed.

Exercise 6.1 Design a memory-based logic system that will enable one decade of 8421 BCD to be converted into any of the following codes: 4221, 7421, 5211, 74-2-1, 84-2-1, -3642 BCD codes, a random BCD code of your choice and Gray code (See Table 1.2 for weighted BCD codes). Only one code conversion is required at any given time, but the user must be able to select any specific code.

Sequential Logic using Memory Elements

A memory circuit can be programmed to perform a sequential logic function provided feedback is incorporated into the system. The address field is partitioned into the external data inputs and the internal states, which are fed back from the output. Figure 6.6 shows a 256 × 4-bit byte-organized memory, configured as a sequential circuit. The memory is programmed via the address and data-in terminals, by compiling an address/contents table from the present/next state and output specifications of the sequential system.

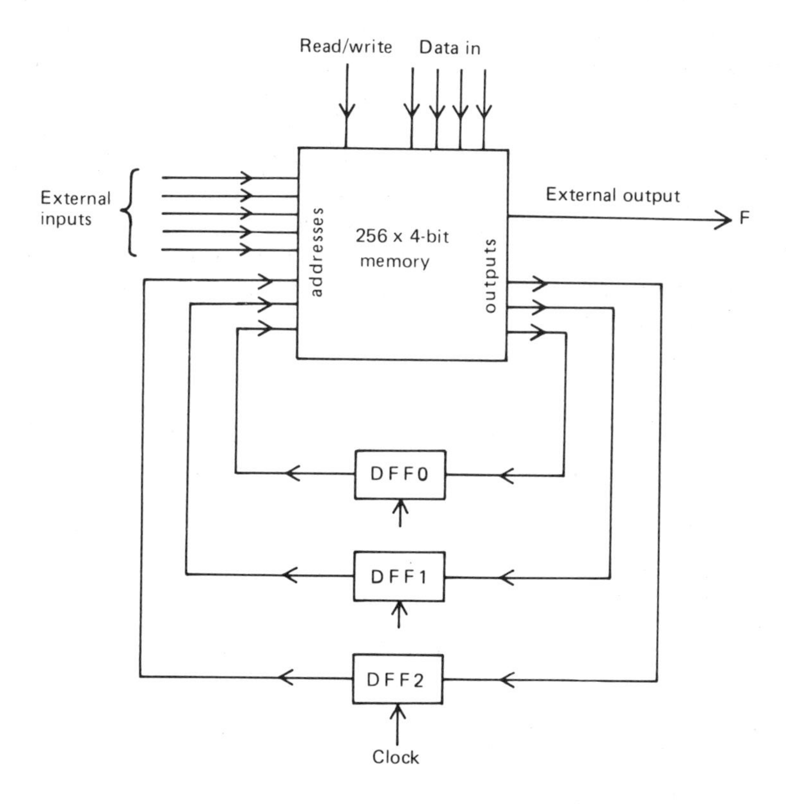

Fig. 6.6 A 256 × 4-bit memory organized as a sequential circuit.

The DFFs in the feedback loop are optional. If these are removed, the circuit becomes asynchronous. The operating time will then be dependent on the access time of the memory, but the system will be susceptible to hazards due to unequal propagation delays in the memory chip. The inclusion of the DFF's synchronizes the feedback and allows the system to switch at a rate determined by the clock driving the flip-flop.

Worked Example 6.4

Program into a memory system, the sequential logic circuit (Worked Example 5.5) which detects the sequence 101 in a stream of bit-serial data.

In Worked Example 5.5 we identified 4 internal states and obtained the present/next state and output tables. They are

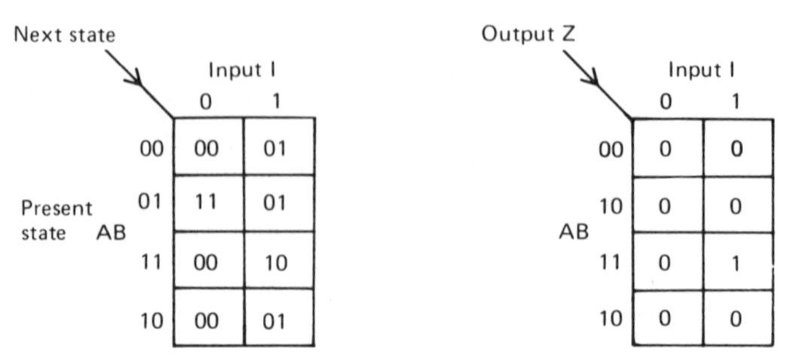

where I is the external input, Z the external output, and A and B the state variables.

The address to the memory must comprise the present state (2 bits) and the external state (1 bit). Hence a 3-bit address field is required. Similarly, the output from the memory provides the logic system's output Z (1 bit) and the next states A′ and B′ (2 bits). The memory needs to be organised in 3-bit bytes giving a minimum requirement of 8 × 3 bits. Let the 3-bit address field be IAB and the 3-bit output be ZA′B′. Address 000 represents I = 0 and AB = 00 and we can see from the sequential system specification tables, that under these input conditions, the output Z needs to be 0 and the next state A′B′ is 00. The 3-bit byte addressed by 000 must therefore be set to 000. By referring to the system tables, the required memory contents for each address can be evaluated.

A′ B′ is the next state, whereas AB is the present state.

They are:

Address	Memory contents
I A B	Z A′ B′
0 0 0	0 0 0
0 0 1	0 1 1
0 1 0	0 0 0
0 1 1	0 0 0
1 0 0	0 0 1
1 0 1	0 0 1
1 1 0	0 0 1
1 1 1	1 1 0

and the block diagram for the memory is:

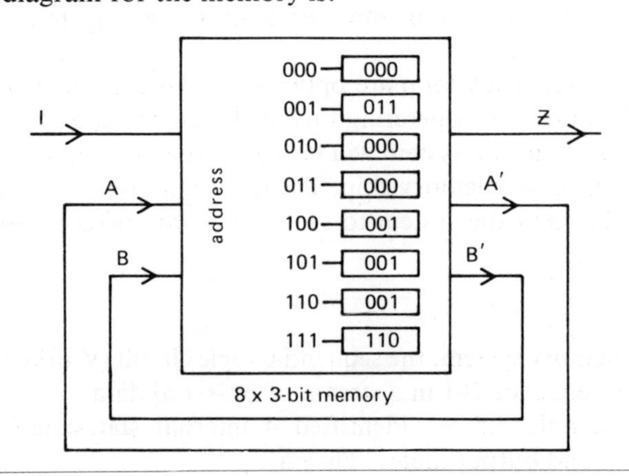

Programmable Logic Arrays

A programmable logic array (PLA) can be regarded as a memory circuit with a programmable decoder. In a conventional memory the decoder is universal. This means that the n address lines can access each and every one of the 2^n storage locations. However, in a PLA, the decoder addresses far fewer than 2^n and, under certain circumstances, this may have significant advantages. Consider the combinational function:

$$F = ABCDEF + \bar{A}\bar{B}\bar{C}\bar{D}\bar{E}\bar{F} \qquad (6.4)$$

A 64-bit memory, having 6 address lines would be required for this function. If however, the decoder could be programmed so that

$$X = ABCDEF$$
and $$Y = \bar{A}\bar{B}\bar{C}\bar{D}\bar{E}\bar{F}$$

and then X and Y used to address the memory, only 4 bits would be required and any input other than the two minterms in Equation 6.4 would address location X Y = 00 and output 0. In a conventional memory these terms would address any one of 62 different cells, all of which must contain 0.

PLAs represent a more efficient way of implementing logic functions and therefore require lower power consumption, fewer and smaller chips for a given system and lower interconnection costs. The electrical behaviour of a PLA is very different from a conventional memory, PLAs do not contain any storage elements as such. They can be regarded as an X.Y matrix of wires and are programmed by making electrical connections between the X and Y intersections. The matrix is divided into 2 parts — the AND Array (equivalent to the decoder) and the OR Array (the store). Data is input via the Y wires into the AND array. There is a wire for each input variable and its inverse. The decoder is programmed by burning connections through to the X wires, which represent the implicants of the logic function, and address the OR array. A programmed logic array is shown in Fig. 6.7. The connections are represented by a heavy dot. The Y wires leaving the OR array represent the outputs of the logic functions.

If a connection is not required, the internal diode at the node is blown. A PLA is a microcircuit and internal connections cannot therefore be made manually.

Implicants are obtained by combining minterms.

An important electrical property of the array is that the connections to the

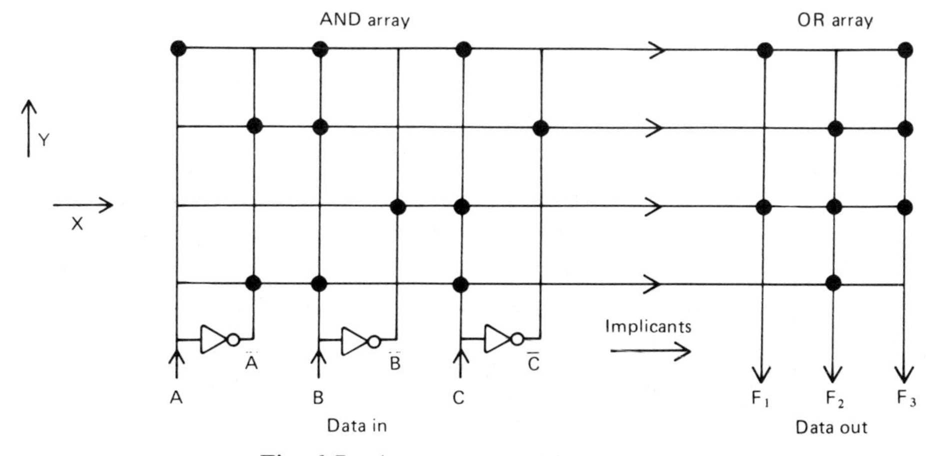

Fig. 6.7 A programmed logic array.

horizontal X wires in the AND array perform the AND function, whereas the connections to the vertical wires in the OR array perform the OR function. The reader should verify that the functions performed by the PLA in Fig. 6.7 are

$$F_1 = ABC + \overline{B}C$$
$$F_2 = \overline{A}B\overline{C} + \overline{B}C + \overline{A}BC$$
$$F_3 = ABC + \overline{A}B\overline{C} + \overline{B}C$$

When using a PLA as a logic system, it must have adequate input and output lines. The third parameter is the number of lines connecting the AND and OR arrays. This represents the total number of different implicants which can occur in the system. A logic system must therefore be minimized so that its set of implicants can be accommodated within the PLA. Common implicants between different output functions within a logic system need only be programmed once into the AND array. The PLA in Fig. 6.7 is a 3 × 4 × 3 array having 3 inputs, 4 implicant lines and 3 outputs.

Use K-maps, Quine – McCluskey algorithm or Boolean algebra. Total minimization may not be needed. The system has to be sufficiently minimal so that the total number of implicants does not exceed the number of X lines.

Worked Example 6.5

Program the following system in a 3 × 3 × 2 PLA

$F_1 = \Sigma(0,4,5)$
and $F_2 = \Sigma(0, 1, 3, 4, 5)$

The system must be reduced to not more than 3 implicants. These can be identified on the K-maps of F_1 and F_2.

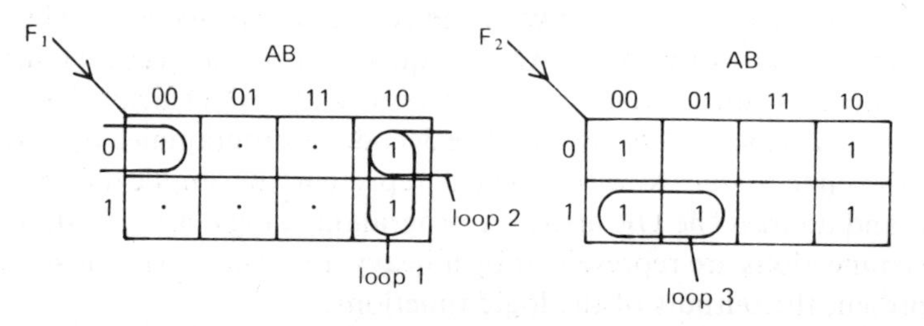

By inspection we can see that loops 1 and 2 are common to F_1 and F_2; the remaining minterms on F_2 are covered by loop 3 giving a total of 3 implicants where

$$F_1 = A\overline{B} + \overline{B}\,\overline{C} \qquad \text{and} \qquad F_2 = A\overline{B} + \overline{B}\,\overline{C} + \overline{A}C$$

F_2 could be minimized further to give $\overline{B} + \overline{A}C$ but this would increase the total number of implicants for the complete system F_1 and F_2 to 4.

The system can now be programmed into a 3 × 3 × 2 PLA. The connections are as follows:

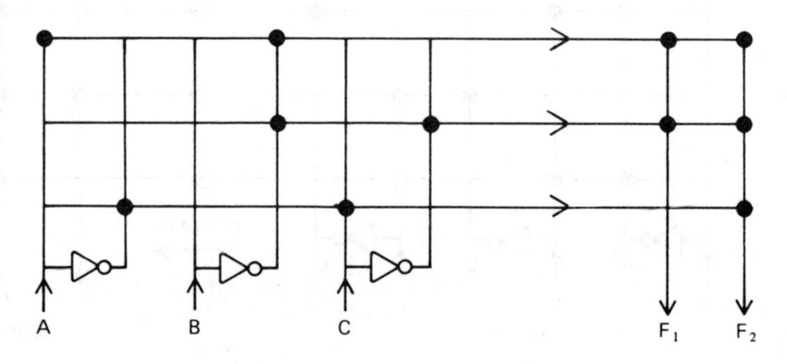

The Microprocessor as a Programmable Logic Device

The digital computer is a product of digital electronic engineering and is now, through developments in solid state technology, available as a single chip microprocessor costing a few dollars. The microprocessor itself can now be regarded as a component in a digital system. It is extremely flexible and can be programmed to perform complex logical and arithmetical operations. Its principal disadvantage is however, its operating speed. Microprocessors are single instruction single data (SISD) devices having an instruction time of the order of 1 microsecond. To perform a complex operation, a program that may run to several hundred operations is needed. As the instructions are carried out serially — one at a time — the execution time may be long, perhaps of the order of fractions of a second. These times are incompatible with very fast parallel logic systems that can operate in a few microseconds. A microprocessor can therefore only be regarded as a viable component of a larger digital system if the desired operations and response times are relatively slow.

SISD — At any instant an SISD machine can only carry out a single operation on one data word.

For a fuller discussion on the microprocessor, refer to Downton, A. *Computers and Microprocessors* (Van Nostrand Reinhold, 1984).

Register Transfers

Truth tables for combinational logic, state tables for sequential logic.

In the designing of digital systems at the circuit level, truth and state tables can be used to specify the behaviour of the system, and we have already developed methods of generating circuits from these specifications. Developments in solid state technology in recent years have, however, made available increasingly complex and cheap integrated circuits, such as adders, counters, timers, memories, encoders and correlators. These and other devices are now becoming the basic components of a digital system.

In order to design a system from complex components, as opposed to gates and flip-flops, we require a means of specifying its behaviour that can then be transformed into interconnections between the components. Register transfer language may be used to represent the order and flow of information between complex components. A register in this context is any self-contained logic system that performs a well defined function on data. The term is not restricted to shift registers. Adders, correlators, timers, counters, etc., can all be regarded as registers.

A typical register transfer statement is

$$T_i\text{: } X \rightarrow A\text{; } Y \rightarrow B$$

T_i is the control data and determines when the operation takes place. It is separated from the rest of the statement by a colon (:) 'X → A' reads 'input X is transferred to store A'. Simultaneous operations at time T_i are separated by a semi-colon (;).

A register transfer statement can define an operation between several registers.

$$T_j\text{: } (A) + (B) \rightarrow A$$

which reads: at time T_j the contents of stores A and B are added together and the sum returned to store A.

To enable the sequence of operations to be carried out in the correct order, a control circuit is required to produce signals to initiate the operations in the order specified by T_i (see Fig. 6.8a). A master clock synchronizes the control circuit and a control unit generates the desired sequence of signals. In complex systems, the

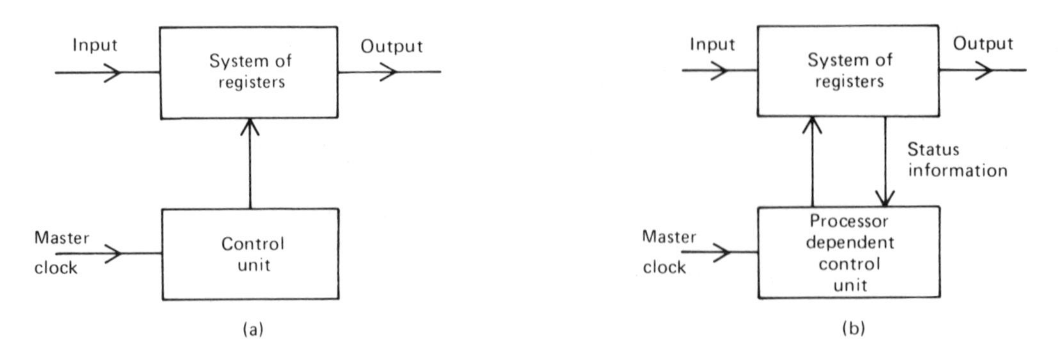

Fig. 6.8 (a) A register controller. (b) A processor dependent register controller.

controller may be dependent on the result of operations in the digital system (see Fig. 6.8b).

Worked Example 6.6 Compile a register transfer specification and design a system to generate 2's complement of an input, using an adder and a store.

2's complement is obtained by inverting the input and adding 1 to the least significant place. We have an adder and a store available, and a register transfer specification for this problem is

T_1: $0 \rightarrow A$ \ clear store
T_2: $X \rightarrow A$ \ input number
T_3: $\overline{A}+1 \rightarrow A$ \ generate 2's complement
T_4: $A \rightarrow$ Output \ output result

Operation T_1 can be achieved by using the clear control on the store. At time T_2 the input is applied to the circuit and the store clocked. T_3 requires the inverse output of the store to be gated into the adder. The carry-in must be set to 1 and the output from the adder returned to the store. Finally, at time T_4 the store contents are gated into the output lines.

The following structure satisfies these requirements:

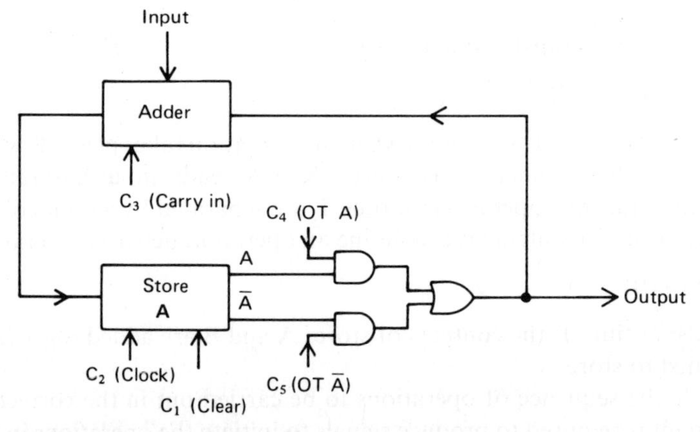

The binary lines represent a parallel word and gating logic is needed in order to output either **A** or $\overline{\mathbf{A}}$ from the store. The 5 control lines are

C_1 — Clear Store
C_2 — Clock Store
C_3 — Set adder carry-in to 1
C_4 — Output **A** from store
C_5 — Output $\overline{\mathbf{A}}$ from store

To generate the 2's complement, the input word is applied to the adder and the control lines set to 1 in the following order

T_1: $C_1 = 1$
T_2: $C_2 = 1$
T_3: $C_4 = 1, C_3 = 1, C_2 = 1$
T_4: $C_5 = 1$

Assume store is edge triggered . . . C_2 can therefore be applied at the same time as C_3 and C_4.

Control lines are 0 at all other times.

The controller can take the form of a 2-bit pure binary counter, driven from a master clock. If the output bits of the counter are B_1 B_0 then

$C_1 = \overline{B}_1 \overline{B}_0$
$C_2 = \overline{B}_1 B_0 + B_1 \overline{B}_0 = B_1 \oplus B_0$
$C_3 = C_4 = B_1 \overline{B}_0$
and $C_5 = B_1 B_0$

The controller circuit is therefore

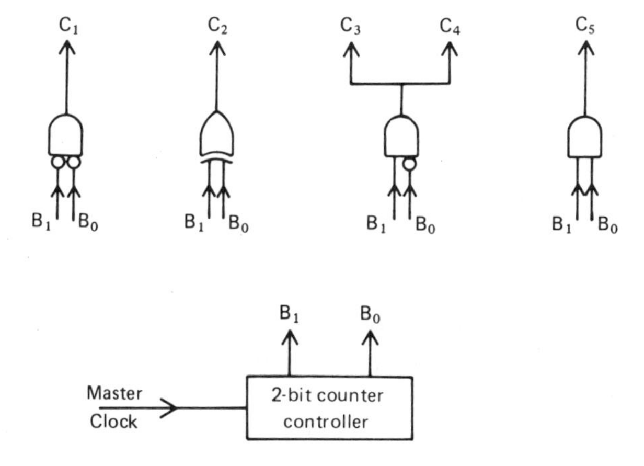

Design an arithmetic unit together with a controller that is capable of adding or subtracting two numbers. The arithmetic unit should be based on a parallel full adder and storage registers.

Exercise 6.2

The Interconnection Problem

When building a digital system from registers, the number of interconnections between modules soon becomes a significant, if not insuperable, problem. Take a system of four 16-bit stores **A,B,C** and **D** where data transfer between any two stores is permissible. A total of 12 data highways as shown in Fig. 6.9 are needed, each containing 16 wires, if the system is to operate in parallel, giving a grand total of 192 wires. This is in addition to any gating logic. Every input line requires an

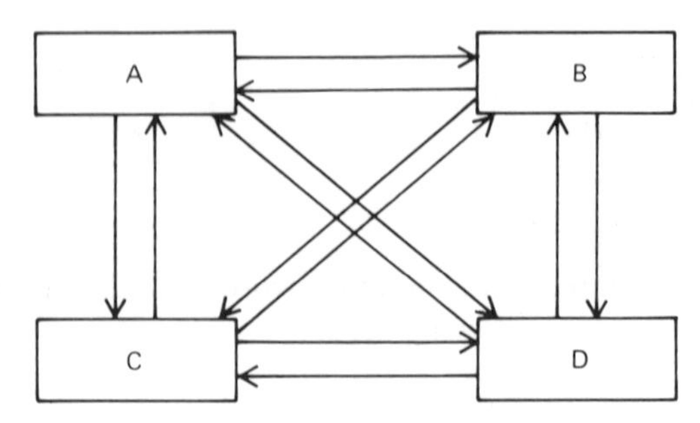

Fig. 6.9 Interconnection of 4 registers.

OR gate to enable inputs from any other register and each output line requires an AND gate so that its output can be selected by a control signal. As more registers are introduced into this system, the number of interconnections increases according to the square law

$$C = (N^2 - N)K \tag{6.4}$$

where C is the number of interconnections, N the registers and K the word length. Hence for a moderately complex system of 16 registers operating on 16-bit words, a conventional interconnection scheme would require 3840 wired links. In most cases this would be regarded as unacceptable. The interconnection problem can be dramatically reduced by using tristate drivers and a data bus.

The Tristate Gate

Some tristate gates are disabled with a logical 1 on the control input. Check with manufacturer's data sheet.

A tristate gate, as its name implies, has three output states. The gate also has an additional control input. When the control input is at logical 1 the gate behaves as an ordinary logic gate. A logical 0 on the control, disables the gate and the output enters its third state, having a high impedance between the earth and power supply rails. In this state, the gate can only supply or sink a few microamps of current and can neither drive nor load any device connected to it. The gate is effectively disconnected from the circuit. Common tristate gates include the NAND, NOR, INVERTER and DRIVER. Tristate flip-flops are also available. The symbols are given in Fig. 6.10.

A driver is a single input (A), single output (F) gate where F=A. It has a high power output and can be used to drive a large number of gates. It is sometimes called a BUFFER. Do not confuse with a buffer store, which simply holds information until required.

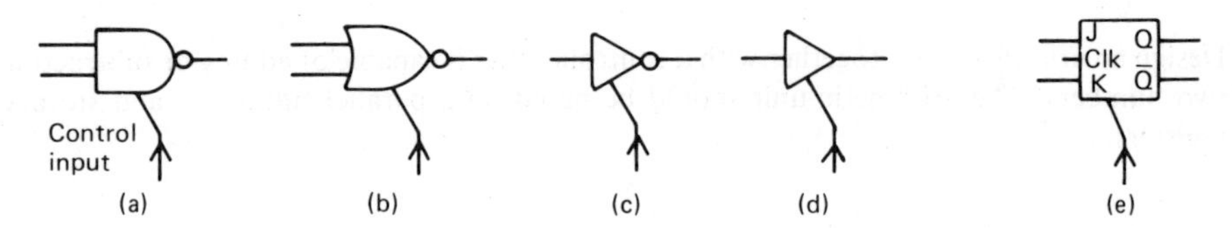

Fig. 6.10 Tristate gates.
(a) NAND. (b) NOR. (c) INVERTER. (d) DRIVER. (e) JKFF.

The Bus Concept

The interconnection problem may be greatly reduced in a register-based digital system by using tristate gates and a bus. A bus is a data highway that runs through-

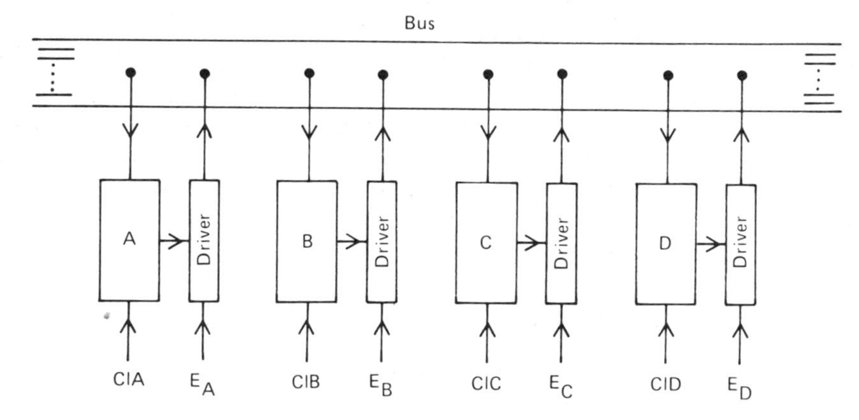

Fig. 6.11 A bus-organized register system.

out a digital system. Connections are made from the bus to the inputs of all the registers in the system, and the outputs are connected via tristate drivers back onto the bus. Fig. 6.11 shows the four-register problem organized on a bus structure.

Registers are often described as being 'hung on the bus'.

When all the control inputs to the tristate drivers are at 0, there is no data on the bus. In order to carry out a data transfer from, say, registers **C** to **A**, the control $\mathbf{E}_c$ to register **C**'s tristate, is set to 1. The contents of that register will be output onto the bus and available at the inputs of all the registers. A pulse is then applied to the clock of the register into which the data is to be written (in this case **A**) and the transfer from **C** to **A** is completed.

Bus systems can be readily extended. Additional registers and drivers may simply be connected to the bus and existing registers do not have to be modified in any way. Almost all large digital systems and computers operate on a bus structure and may have more than one bus if high speed operation is required.

Fault Diagnosis

An integrated circuit may contain tens of thousands of electronic elements, each of which can develop faults. It is also very difficult to recognise most faults in a digital system. If a fault develops in an analogue circuit, for example, a television set, it will result in distortion and perhaps loss of picture or sound. It is generally quite obvious that something is wrong. However, a failure in a digital system is likely to result in a single bit on an output having the wrong value, a 0 instead of a 1 or vice versa. The output will be a valid binary number but it will have the wrong value. It is therefore essential to test digital systems during their operational life for faults that cannot be observed in the output data.

Testing involves applying a set of inputs to the circuit and knowing the outputs that should be generated. The actual outputs are compared with the expected outputs and if they differ, a fault is present. If we examine a combinational circuit having n inputs, there are a total of 2^n possible input states each of which has a well-defined output. The system must be tested on every input state in order to ascertain if any faults are present. This exhaustive testing amounts to checking the circuit against its truth table and is viable provided n is small. In larger systems exhaustive testing will take up too much time.

lk = 2^{10} = 1024 bits, therefore 64k = 65 536 bits.

Consider a 64k-bit memory. It has 16 address lines and 65 536 storage locations. A total of 131 072 tests are required to check that a 1 or a 0 can be stored in each location. Unfortunately however, this may not reveal all the faults in the circuit. The fault could be dependent on the contents of the memory and only occur when a particular pair of locations are set to 1. There are ${}^{65536}C_2$ or approximately 2×10^9 different pairs of locations which can be set to 1. Scanning the memory for each store setting, assuming 1 microsecond per read operation, would take

$$65\,536 \times 2 \times 10^9 \times 10^{-6} \text{ seconds}$$

or nearly 100 years. Clearly exhaustive testing is impractical except for very simple systems.

In practical fault detection, certain assumptions are made. Only 'stuck' faults are considered. These are the most commonly occurring faults, owing to the elements in the circuit becoming open or short-circuited and result in data lines being stuck at either logical 0 or 1.

Furthermore, only one fault is assumed to occur at a given time, and diagnosis is assumed to be sufficiently frequent to enable a fault to be rectified before the next one occurs.

The combinational circuit that generates the carry in a single bit full adder is shown in Fig. 6.12. Its equation is:

$$F = A\,B + A\,C + B\,C \tag{6.5}$$

Stuck faults can occur on the data lines a to j, and the aim of a fault-assessment procedure is to determine a set of non-exhaustive tests that will reveal 'stuck' faults on these lines.

Suppose a fault causes g to be stuck at 0. The function will become

$$F(g) = A\,C + B\,C \tag{6.6}$$

In order to detect this fault we require an input that will either give $F = 1$ and $F(g) = 0$ or $F = 0$ and $F(g) = 1$. If the test data produces the same output from the faulty and fault-free circuits, then the fault will not be detected. Therefore the EXCLUSIVE-OR of F and F(g) must have a value of 1:

$$F \oplus F(g) = 1 \tag{6.7}$$

By substituting for F and F(g) in Equation 6.7, a suitable test can be obtained:

$$(A\,B + A\,C + B\,C) \oplus (A\,C + B\,C) = 1$$

gives $A\,B\,\overline{C} = 1$ (6.8)

The fault can therefore be detected by A B C = (110). This input when applied to the fault-free circuit gives an output of 1, but if line g is stuck at 0, the output becomes 0.

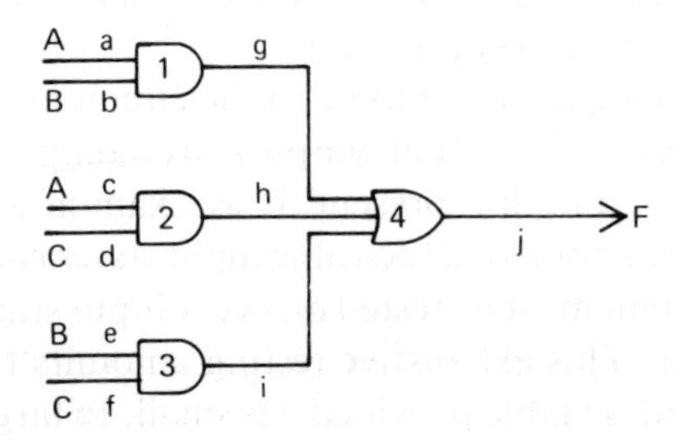

Fig. 6.12 The carry circuit in a full adder.

Exercise 6.3

In the following circuit, the output of gate 3 is stuck at 1. Devise a test to identify this fault.

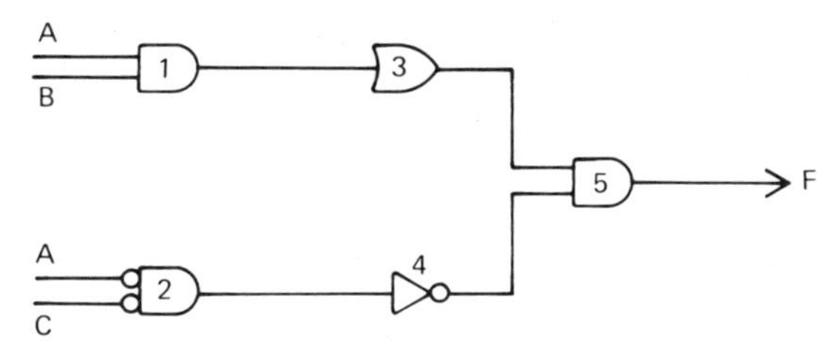

Indistinguishable and Undetectable Faults

If a circuit is in its minimal form, all individual 'stuck' faults can be detected. Unfortunately, there is not necessarily a unique test for every fault. Different faults may be detected by the same test and it may be impossible to distinguish between them.

Consider the carry generator shown in Fig. 6.12. Suppose input line c is stuck at 1. The equation of the faulty circuit becomes

$$F(c) = AB + C + BC \tag{6.9}$$

The test that detects this fault is given by

$$F \oplus F(c) = 1$$

$$\therefore (AB + AC + BC) \oplus (AB + C + BC) = 1 \tag{6.10}$$

giving $\overline{A}\,\overline{B}C = 1$

Equation 6.10 need not be evaluated using algebra. Compare K-maps to see which input values set F to 1 and F_c to 0 or vice versa.

The fault will therefore be detected by input state

$$ABC = (001)$$

Now, if line e is stuck at 1, the equation of the circuit containing the new fault becomes

$$F_e = AB + AC + C$$

and $F_e \oplus F = 1$

also gives $ABC = (001)$

Hence e stuck-at-1, requires the same test as c stuck-at-1. ABC = 001 will detect at least two different faults.

In order to determine whether a test exists that will distinguish between the two faults, the EX.OR of the functions of the two faulty circuits is calculated. If it is zero for all inputs, then the faults are indistinguishable.

So $F_e \oplus F_c$

$$= (AB + C + BC) \oplus AB + AC + C \tag{6.11}$$

By EX.ORing corresponding cells on the K-maps of F_e and F_c we obtain

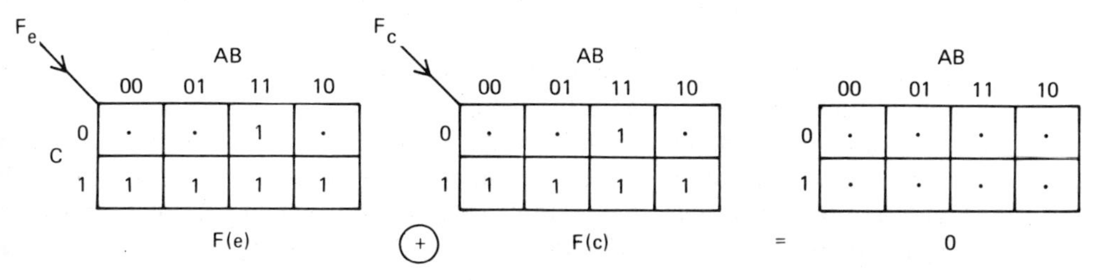

$F_e \oplus F_c$ is zero for all inputs, and the faults are therefore indistinguishable.

In non-minimal circuits, it may be impossible to detect some faults. Therefore automatic fault detection is envisaged, a system must be designed in its minin form.

Worked Example 6.7

Show that it is impossible to detect a 'stuck-at-0' fault on the output of gate 3 in t following circuit:

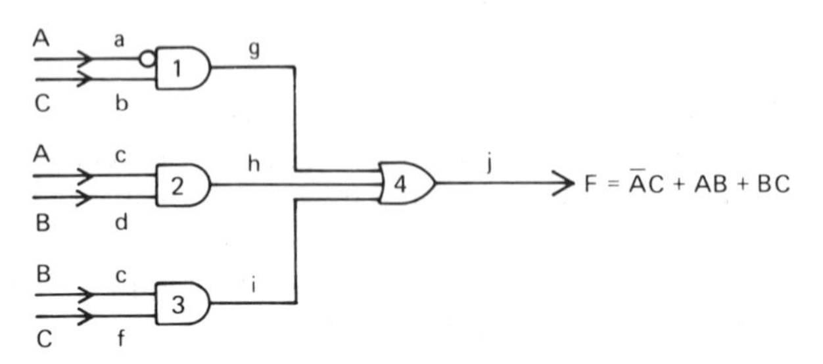

The equation of the faulty circuit is

Evaluate truth tables for F and F(i) — they are identical.

$$F(i) = \overline{A}C + AB + 0 = \overline{A}C + AB$$

A test to detect i 'stuck-at-0' must satisfy

$$F \oplus F(i) = 1$$

But $(\overline{A}C + AB + BC) \oplus (\overline{A}C + AB) = 0$

$\therefore$ No test will detect fault i 'stuck-at-0'.

For further details on Fault Diagnosis refer to Breuer, M.A. and Friedman, A.D. *Diagnosis and Reliable Design of Digital Systems* (Pitman, 1977).

Fault diagnosis in digital circuits is the subject of much current research. It now viable for combinational logic systems; however, the problem of faults sequential logic is considerably more complex, and effective methods have yet to discovered.

Fault-Tolerant Systems

The ultimate logic system is one that is tolerant to faults and does not malfuncti when a fault occurs. This is usually achieved by having massive redundancy. some of the systems used in space vehicles, as many as five processors are used perform the same operations. The outputs are correlated and a majority decisi made on the results. In this way, up to two systems can fail without corrupting t outputs.

At a less sophisticated level, however, redundancy still provides a tolerance faults. In Worked Example 6.7, the 'stuck-at-0' fault on line i of gate 3 does r corrupt the output of the circuit. The faulty circuit has the same K-map as t original function, because gate 3 is redundant.

Thus the digital designer is left with a dilemma. Either he provides a minin design that is fault diagnosable and can easily be repaired, or he must opt fo highly redundant system that is fault tolerant, but which cannot be fully fa diagnosed and is therefore difficult to service. One would expect the former to ha a relatively large number of short periods when non-operational, whereas the latter will give a prolonged period of operation until the number of faults is sufficiently large that they cannot be overcome by the redundancy. An exhaustive service operation would then be required.

Summary

Programmable logic can perform any desired function without having to make any physical changes to the hardware structure. A multiplexer may operate as a universal combinational logic function. It is programmed directly from a truth table by means of external logic switches on the data-in-lines. Multilayer multiplexer logic systems can be devised for larger logic functions. The memory element may also act as a programmable logic function. As the truth table is stored internally, a memory-based logic function does not require the 2^n input connections needed for an n-variabled function implanted on a multiplexer. The truth table can be entered serially via a single input line.

A PLA can be regarded as a memory circuit with a programmable decoder. Its structure provides a more efficient means of implementing logic functions in terms of hardware size and power consumption. If memories or multiplexers are used, a function need only be minimized as far as the removal of redundant input variables, whereas with PLAs the system must be reduced to the number of implicants the PLA is capable of accommodating.

The incorporation of feedback within a programmable logic structure, enables sequential logic system to be implemented.

Register transfer language is used to specify the operation of a system based on large scale integrated circuit components (the registers). As the number of registers in a system increases, the number of interconnections becomes unacceptably large. The interconnection problem can be greatly reduced by using tristate gates and a bus structure.

Fault diagnosis of large and complex systems is becoming an increasingly important subject. Test pattern generation based on EXCLUSIVE-OR behaviour between faulty and fault-free systems has been examined and finally the possibility of fault tolerant systems was proposed, by allowing redundancy in the system.

Problems

6.1 Determine the input settings required on an 8 to 1 multiplexer if it is to implement the following 3-variabled functions:

(i) $F = \Sigma(2,4,5,7)$
(ii) $F = \Pi(2,5,6)$
(iii) $F = \overline{AB + \overline{A(B+C)}}$

6.2 Determine the input settings required on an 8 to 1 multiplexer if it is to implement the following functions of 4 variables:

(i) $F = \Sigma(1,3,5,6,9,12,14)$
(ii) $F = \Pi(0,1,2,3,4,5,6,7,12,13)$

6.3 Design a two layer system of 4 to 1 multiplexers to implement the function

$$F = ABC + \overline{A}\overline{B}D + D\overline{E} + AE$$

6.4 An electronic voting system has 4 voting stations where votes for, against or abstensions can be registered. Devise a suitable logic system using a memory implementation, to indicate when a motion is accepted.

6.5 Design a feedback memory circuit to measure, in Gray code, the duration of a pulse, to 5 bits resolution.

6.6 What size PLA is required to implement the following 3-function system?

$$F_1 = A\overline{B} + \overline{A}$$
$$F_2 = \overline{AD} + ABC$$
$$F_3 = A\overline{C}D + \overline{A}\overline{B}\overline{D}$$

Specify the connections in the array.

6.7 Implement the system:

$$F_1 = \overline{A}BC + AB$$
$$F_2 = (A + B)(A + C)$$
$$F_3 = BC + A\overline{B}$$

in a 3 × 3 × 2 PLA.

6.8 Obtain a register transfer specification of a hardware multiplier which uses a successive addition procedure.

6.9 Examine the following circuit and determine which tests, if any, can identify a stuck-at-1 fault on any of the lines a to h.

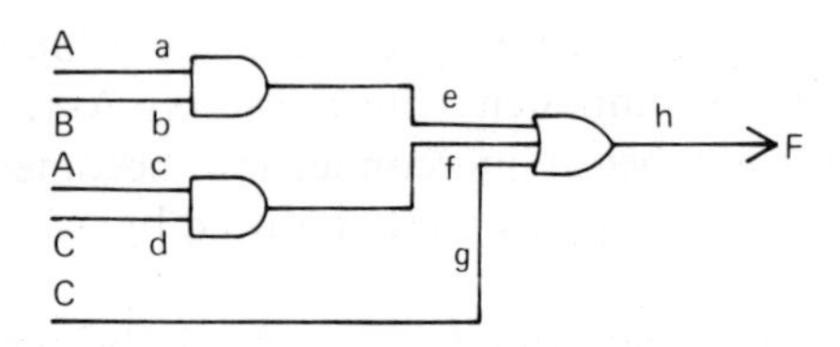

6.10 Repeat Problem 6.9 for a stuck-at-0 fault on any one of the lines a to h.

Practical Digital Circuits 7

Objectives

- ☐ To examine and compare different logic functions.
- ☐ To identify constraints on the use of logic components.
- ☐ To investigate methods of interconnecting different logic families.
- ☐ To survey practical memory circuits.
- ☐ To design converters to interface the digital system with the analogue world.
- ☐ To define and eliminate potential hardware problems.

In this final chapter, we shall concentrate on some of the practical aspects of a digital logic system. The different logic families will be identified and their characteristics examined. Practical constraints that must be observed when a logic system is put together, will be discussed. Furthermore, a working digital system does not exist on its own. It has to be interfaced with other systems, such as industrial processes, machines and people. Interfacing the digital computer with the analogue world is therefore an important area, and the design of data converters for this purpose will be outlined.

Logic Families

Integrated circuit logic families can be divided into two broad groups: the bipolar and the metal oxide semiconductor (MOS) families. Both rely on the switching of transistors between two discrete states in order to represent the logical behaviour of a function. The bipolar logic families are based on the bipolar transistor, whereas the MOS systems use the unipolar field effect transistor as their component. Both types are suited to integrated circuit manufacture.

The two states are usually cut-off and saturation.

Bipolar Transistor Logic

The earliest electronic logic was based on diodes that could represent the logical constants by being either forward or reverse biased. Diode logic circuits cannot, however, perform the NOT operation, and there is a loss of voltage at each gate, owing to the contact potential of the diode. Therefore, if a circuit has several gates following one another, it is possible that the voltages representing the logical constants 0 and 1, could overlap and the system fail. The transistor can also act as a two-state switch, and as an amplifier with 180° phase change. All these properties are needed in a logic circuit. The two distinct states represent the logic constants, and amplifying property enables the voltage levels representing the states, to be maintained, and the 180° phase shift acts as an inverting function.

Contact potential approx. 0.3V for germanium and 0.7V for silicon diodes.

Resistor Transistor Logic

The earliest form of transistor logic was resistor transistor logic (RTL) where input

Beware. RTL also stands for 'register transfer language'.

variables were applied via biasing resistors to the base of a transistor. The values of the resistors were chosen, so that the inputs would, depending on their voltage levels, either saturate or cut-off the transistor.

Diode Transistor Logic

Diode transistor logic (DTL) where the logical AND or OR operations were performed by the diodes and the transistor acted as a voltage amplifier and inverter, appeared in the early 1960s. Compared with RTL, DTL offered faster switching speed, as the diode input circuits provided lower driving resistance to transistors.

Transistor Transistor Logic

In 1964, Texas Instruments introduced their 7400 series of transistor transistor logic (TTL), which has dominated the logic component market for the past two decades and is still widely used today. The principal features of a TTL gate is a multiple emitter input transistor that is functionally equivalent to the input diodes in a DTL circuit.

Since its inception, the TTL logic structure has been continuously developed to meet more stringent speed and power consumption standards. The following TTL series are currently available:

74HOO Series

(i) High speed TTL where the circuitry is basically the same as the standard TTL but the resistors within the circuits have been reduced in value, resulting in faster switching at the expense of increased power consumption.

74LOO Series

(ii) Low power TTL where the resistor values have been increased giving reduced power consumption at the expense of longer propagation times.

74SOO Series

(iii) Schottky TTL which has a Schottky barrier diode connected between the base and collector of every transistor in the gates. A Schottky barrier diode has a contact potential of about 0.25V, compared with 0.7V for a silicon diode, and prevents the transistor saturating heavily. It can therefore, change state much faster and is the highest speed TTL available.

74LSOO Series

(iv) Low power Schottky TTL, where the use of Schottky diodes and high resistor values lead to low power consumption, but the reduction in speed is less than in the low power range, owing to the diode clamps.

Emitter Coupled Logic

Sometimes known as 'current mode logic' (CML).

In the standard TTL ranges, the transistors are saturated during part of their operating cycle. If the saturation is restricted, as in the case of Schottky TTL, the switching speed is increased, resulting in faster devices. An alternative non-saturating bipolar family known as emitter coupled logic (ECL) is available, where the switching is controlled by current generators. When the transistor is switched from the cut-off to conducting state, the resulting collector current is controlled by the fixed current injected into the base which is less than the saturation current I_c (sat). Consequently, ECL provides very high speed switching, being marginally faster than Schottky TTL and having approximately the same power consumption.

ECL gates have different voltage levels for the logic states (−0.8V and −1.7V) compared with TTL (0V to 0.4V and 2.4V to 5.0V) and are not therefore directly compatible. An ECL gate, however, usually generates both normal and inverted outputs and thereby eliminates the need for inverters.

Integrated Injection Logic

Integrated injection logic (I^2L) is the latest bipolar family. It offers a range of speed/power consumption trade-offs and can operate in both saturated and unsaturated modes. High packing density is achieved because biasing resistors are external components not included within the chips. It is likely therefore, that I^2L will have a significant effect on VLSI design.

A resistor within an integrated circuit takes up 10 times more chip area than a transistor.

The power/speed characteristics of bipolar transistor logic families are summarized in Table 7.1.

Table 7.1 Typical Power and Propagation Data for Bipolar Logic Families

Name	Series number	Propagation (nsec)	Power (mW)
Standard TTL	7400	10	10
High speed TTL	74H00	5	25
Low power TTL	74L00	35	1
Schottky TTL	74S00	3	25
Low power Schottky TTL	74LS00	10	2
Emitter coupled logic	ECL	2	25
I^2L lowest power		100	5×10^{-6}
I^2L fastest speed		5	5

Unipolar Logic Families

Unipolar logic families are based on the field effect transistor which requires a metal electrode separated from a semiconductor channel by an oxide insulating layer. This is the MOS fabrication technology, and the individual transistors are often referred to as MOSFETS.

MOSFETS: metal oxide silicon field effect transistors

MOS logic devices, unlike bipolar logic (except for I^2L), do not require internal resistors on the chips, and can hence be manufactured at a high packing density. The fabrication process is relatively simple, and power consumption is low. MOS ICs however have relatively slow operating times and are therefore not suited to logic systems where very fast switching is essential.

Ideal for battery-operated systems

P-MOS Logic

P-MOS logic is a family of MOS integrated circuits. The semiconductor material is p-doped and the majority carriers are therefore holes. P-MOS is suited to large scale integration and has a greater packing density capability than bipolar transistor logic.

N-MOS Logic

N-MOS Logic is basically the same structure as P-MOS, except that it uses n-doped semiconductor material. The circuit carriers in N-MOS are free electrons, which are more mobile than larger positive charges, and this results in faster switching times compared with P-MOS devices.

CMOS Logic

Complementary metal oxide semiconductor logic uses both P and N type channels in the same circuit. It is faster than PMOS and NMOS and requires considerably less power than the low power TTL series. CMOS is however still inferior to standard TTL in terms of operating speed. MOS circuits can operate off a wide range of supply voltages. If CMOS is being used alongside TTL, a 5V supply would be used. However, when used alone, it will operate satisfactorily over the range 3 to 20V. The input levels are dependent on the supply voltage V_s and are approximately 30% and 70% of V_s for the low and high states, respectively. Power dissipation is also dependent on the supply voltage, but is still extremely low compared with other families.

Silicon on Sapphire

For electronic circuit analysis of bipolar and unipolar transistor logic gates refer to Chapters 6 and 7 of Richie, G.J. *Transistor Circuit Techniques — discrete and integrated* (Van Nostrand, Reinhold, 1983).

SOS is a recent development in MOS technology where the oxide insulating materials is replaced by sapphire. This has the effect of reducing internal capacitance and thereby increases the speed of the logic.

Practical Constraints on Logic Circuits

All logic families are subject to certain constraints in use. The main ones are identified below, but the designer must always familiarize him/herself with the manufacturer's data sheet when using particular products.

Fan-out

Special drivers are available that will drive more than 10 gates.

Fan-out is the number of logic inputs that can be driven from the output of a logic gate. A typical fan-out for a TTL gate is 10. However, with CMOS logic, owing to the very high input impedance of the field effect transistor ($10^{11}\Omega$), the fan-out is potentially infinite. In practice, however, owing to capacitive loading of the driving output by the inputs, the fan-out is generally restricted to a maximum of 50.

Fan-in

VERY IMPORTANT

Fan-in is the number of logic outputs that can be connected to a logic gate input. A fan-in greater than 1 was possible with the now obsolete RTL, where every output was fed via its own external resistor, to the input in question. Standard TTL outputs, however, must NEVER be connected together, as the internal transistors will be overloaded if one output is low and another is high. The fan-in of a standard TTL gate is therefore 1. Special open collector gates can be obtained, where the output transistor load is not included on the chip. It has to be supplied externally. The outputs of open collector gates can be connected together and a 'wired-AND' operation obtained. The output across the external load resistor is high only if all the gate outputs are high. If one or more gate inputs are low then the 'wired-AND' output goes low.

Unused Inputs

For example, to obtain a 2-input NAND function from a 3-input gate, the third input must be held at logical 1.

If one or more inputs to a gate are not needed for a particular logic function, the unused inputs should either be connected to logical 1 or logical 0, depending on the function being performed. Alternatively, they may be connected in common with a used input line to form a single input. Unused inputs must NOT be left unconnected, as they are highly susceptible to noise and pick-up which can give rise to incorrect logic behaviour.

An unconnected TTL input, in the absence of noise or pick-up will appear to be high (logical 1 with positive logic coding) but in a practical circuit this cannot be guaranteed. In CMOS circuits a floating input could bias both P and N channels towards their conducting states and lead to excessive power consumption.

Handling Precautions

A rather surprising problem arises with CMOS devices, which can result in the circuits being damaged unless suitable precautions are taken. Owing to the extremely high input impedance that can be tens of thousands of megohms, static charge can build up on the input terminals of a CMOS chip, which can lead to breakdown within the circuit. Great care should be taken when handling CMOS components, and the devices should always be stored with their terminals embedded in conducting foam, to prevent this static build-up.

Interfacing CMOS and TTL

In many applications the designer may wish to use CMOS logic, where low power requirements are a priority, and TTL for the parts of the system working at high speed. A system may therefore contain both CMOS and TTL families. The supply rail must be +5V for TTL, whereas CMOS requires a supply voltage of between 3V and 20V. A common +5V rail will therefore satisfy both families.

CMOS Output driving TTL input

A high output from a CMOS gate can drive a TTL input, but when the CMOS is low, the input current that flows back from the TTL gate through the CMOS output to earth (see Fig. 7.1), is sufficiently large to cause a potential drop across the CMOS output resistance that will appear as a logical high to the TTL input. A typical value for the output resistance of a CMOS gate is 2kΩ and the current drain

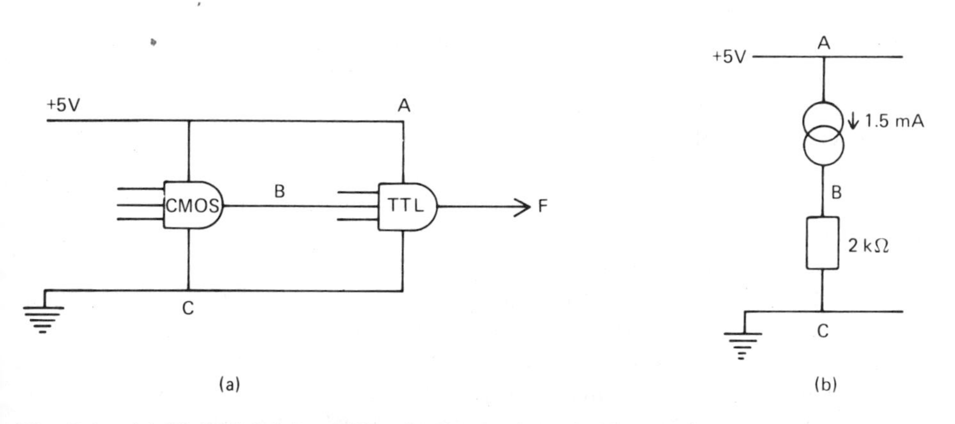

Fig. 7.1 (a) CMOS driving TTL. (b) Equivalent circuit of path A → B → C in (a). Point B cannot fall below 3.0V.

of a TTL input, 1.5mA, giving a voltage of the order of 3V at the TTL input, which would appear to be a logical high. Standard CMOS cannot therefore drive standard TTL. A special CMOS buffer must be used that is designed to sink the current flowing back from the TTL input. CMOS can, however, drive the low power TTL families as the input currents are much smaller compared with the standard series.

Low power (74LOO) and low power Schottky (74LSOO)

TTL Output driving CMOS inputs

TTL will drive CMOS because its low output is approximately 0.4V on open circuit and this is within the range 0 – 30% of V_s (0 – 1.5V) acceptable as a low input for CMOS. The open circuit high output for TTL is about 3.6V which is just within the range of > 70% of V_s (3.5 – 5.0V) acceptable as a high input to CMOS. The TTL output is however dangerously close to the lower limit of the range and a mere 0.1V noise could cause a malfunction. In practice, therefore, external pull-up resistors are used to increase the high output voltage of the TTL gate and hence give greater reliability (Fig. 7.2).

CMOS input impedance is extremely high and therefore appears as an open circuit to the TTL.

$V_s = +5V$

Typical resistor value is 1kΩ

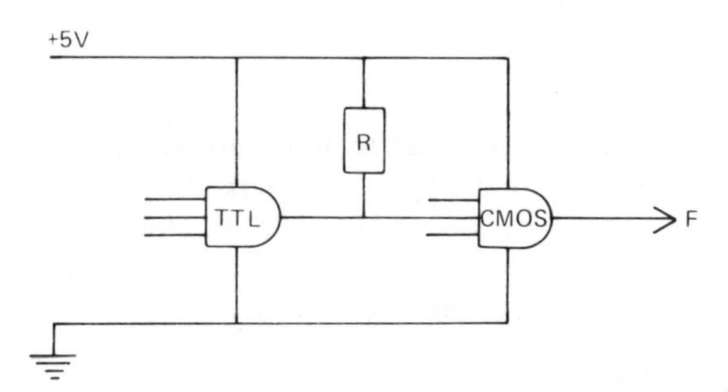

Fig. 7.2 A pull-up resistor used when TTL drives CMOS.

Practical Memory Circuits

Memory devices have two very important roles in logic design. They are storage devices and can also act as programmable universal logic functions. Consequently memory technology has received much attention during recent years. Memories can be divided into two broad groups — the random access memory or RAM, and the read only memory or ROM.

The RAM is a volatile memory. Its contents are lost when the power supply is turned off. The time taken to read the contents of a storage location is of the same order as the time needed to write information into the store. Two types of RAM are available — static and dynamic. In the static RAM, storage elements are flip-flops and the stored data remains constant, provided power is supplied to the chip. The dynamic RAM relies on stored charge on capacitors within the integrated circuit. Stored charge represents logical 1, whereas no charge indicates logical 0. Unfortunately, capacitors will not hold charge indefinitely. It will gradually leak away. Consequently a dynamic memory must be read periodically and those capacitors storing logical 1 must be recharged. This process is called 'refreshing' and must be carried out every few milliseconds. Nowadays, dynamic RAMs have refresh logic

Compared with a read or write time of 100 nanoseconds.

included on the memory chips and no additional external circuitry is required. Dynamic RAMs require less circuitry per bit than static RAMs. Approximately four times more memory can be fabricated on a given area of silicon chip, if dynamic rather than static storage is used.

The ROM is, by contrast, a non-volatile memory. The stored data is NOT lost when the power supply is turned off. The write procedure is very much longer than the read operation. Write time to read time ratios may be as high as 10^9 for the ROMs that can be programmed by the user.

ROM: read only memory.

It may take several minutes to write into a ROM.

The simple ROM consists of an XY matrix of wires. The X wires are addressed by a decoder and the Y wires provide output data as described in Chapter 6. The data is stored by making connections between the XY intersection during manufacture.

To store a 1, connect X to Y. To store a 0, leave XY open circuit.

Programmable ROMs or PROMs have a coupling device in series with a fusible link at each XY intersection. To program the device, the user must pass a large current through each link where a logical 0 needs to be stored. The link is thereby burned or open-circuited, causing the output to go low.

A diode or a transistor.

EPROMs are erasable PROMs. The conduction paths and hence the stored data in these memories are defined by the build-up of charge within the gates. This charge can be released and the memory erased if the device is exposed to high intensity ultraviolet radiation for several minutes. The device can then be reprogrammed.

The most recent addition to the memory range is the EAROM or electrically alterable ROM. It is user-programmable and electrically erasable, and thereby dispenses with the ultraviolet radiation source required by the EPROM.

For further details on memories refer to Chirlian, PM. *Analysis and Design of Integrated Electronic Circuits* Vol. 2 (Harper & Row; 1984).

Interfacing the Digital System to the Real World

We saw in Chapter 1, that the real world invariably appears to be continuous or analogue, whereas operations in a digital system are, by definition discrete. In any practical application of a digital system, an interface with continuous or analogue processes is needed. This interface is achieved by using analogue to digital converters (ADC) at the input and digital to analogue converters (DAC) at the output of a digital system.

Digital to Analogue Converter

A simple DAC can be built up from a series of weighted resistors. Taking a 4-bit binary word

$$\mathbf{B} = b_3\, b_2\, b_1\, b_0$$

the most significant bit is connected to a resistor of value $2R$, bit b_2 to a $4R$ resistor, b_1 to an $8R$ resistor and so on. The value of the resistor doubles for each less significant bit. If any bit is at logical 1, a voltage is applied across its resistor, and a current inversely proportional to the value of the resistor flows. An operational amplifier circuit can be used to sum the currents and produce a voltage proportional to the total current flowing. The magnitude of the voltage will therefore be directly proportional to the value of the binary number **B**. The block diagram of a weighted resistor DAC and its circuit diagram, using an operational amplifier are

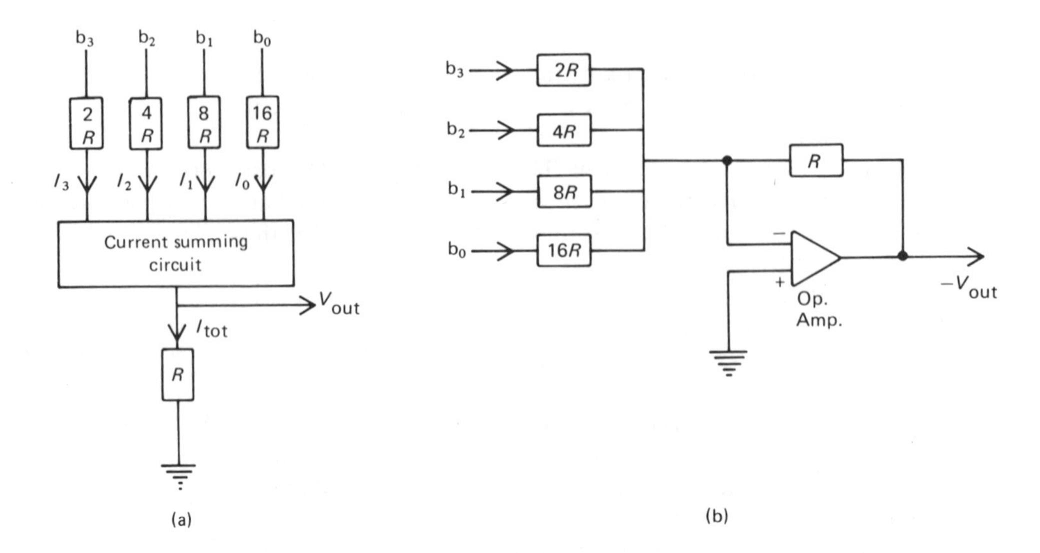

Fig. 7.3 A weighted resistor DAC. (a) Block diagram. (b) Circuit diagram using an operational amplifier.

shown in Fig. 7.3. The weighted resistor DAC, although very simple, has two major drawbacks: a wide range of resistor values is needed, and the ratio of the smallest to largest resistor is an exponential function of the number of bits.

The resolution is $n + 1$ bits

$$\frac{R_0}{R_n} = 2^n \tag{7.1}$$

So, if a 16-bit converter was constructed and the resistor for bit b_{15} was 2kΩ, the least significant bit resistor would need to be $2^{15} \times$ 2kΩ or 65.536MΩ. Furthermore, the resistors must be sufficiently accurate, so that the contribution to the total current from the least significant bit, is not masked by the current arising from inaccuracies in the most significant bit resistor. For a 16-bit converter the accuracy of the resistor R_{15} must be better than 0.0015% and, in practice, it is impossible to manufacture resistors to such a high accuracy over a wide range of values. Hence weighted resistor DACs have limited practical use, except for very low resolution conversions.

Exercise 7.1 Design a 4-bit weighted resistor DAC and calculate the maximum tolerance on the resistors if the converter is to be accurate to the nearest least significant bit.

An alternative approach to digital to analogue conversion is found in an $R/2R$ resistor ladder network as shown in Fig. 7.4. This converter only requires two different resistor values and the network can be fabricated as an integrated circuit. Every bit in the binary word **B** controls a switch. The most significant bit is applied to SW_3 through to the least significant bit controlling SW_0.

The resistor network is a current dividing circuit and each switch directs the current in its branch either to earth if the input driving the switch is at logical 0 or to the current summing operational amplifier if at logical 1. The operational amplifier input (node F) appears to be an earth as far as the resistor network is concerned. Hence the network can be analysed by assuming that the $2R$ resistors are connected to earth, regardless of the switch settings.

The input of an inverting operational amplifier is a virtual earth. See Horrocks, D.H. *Feedback Circuits and Op. Amps.* (Van Nostrand, Reinhold, 1983).

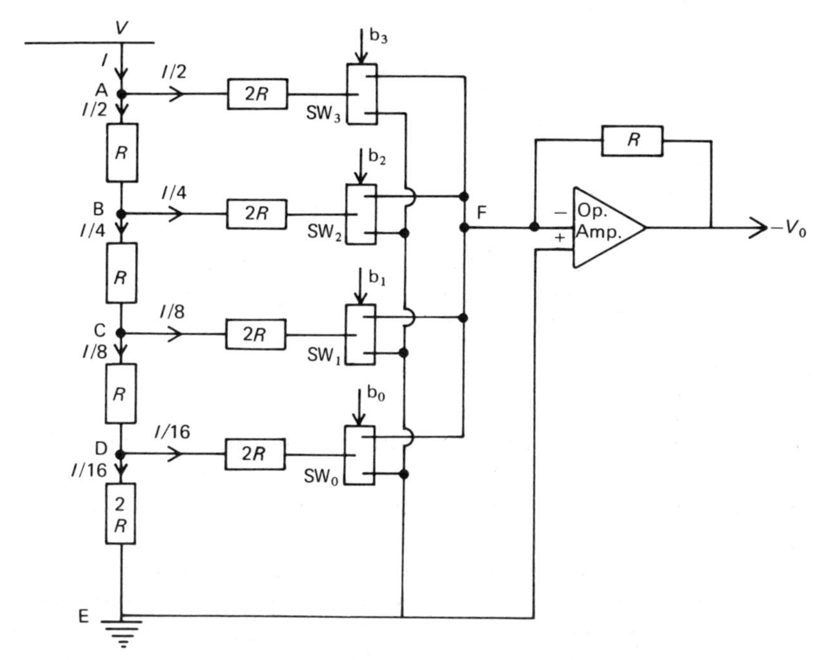

Fig. 7.4 An $R/2R$ resistor ladder network digital to analogue converter.

At each node (A,B,C,D) in the ladder network, the resistance is $2R$ when looking either towards the switches or towards earth point E. Hence any current flowing into the node will be divided equally. Only half will flow towards the switch and the other half flows towards the next node, where it is again divided. If the current flowing from the power supply is I, then $I/2$ flows to SW_3, $I/4$ to SW_2, $I/8$ to SW_1, and so on. The digital word **B**, which controls the setting of the switches, will determine whether these current contributions are summed or returned to earth. Hence the total current entering the summer is determined by the digital input and appears as a proportional analogue voltage at the output. The resolution of the converter can be increased by extending the $R/2R$ ladder network.

Analogue to Digital Converters

One of the most widely used analogue to digital conversion techniques is the successive approximation method, where a voltage is generated by a digital to analogue converter and continuously compared with the analogue input. When the voltages are equal, the digital equivalent of the analogue input is the value of the final input to the DAC.

An analogue input is always a voltage.

A block diagram of the successive approximation ADC is given in Fig. 7.5. The controller can be a simple cyclic binary counter and a digital output occurs every time $V_D = V_A$. This method is relatively slow, requiring 2^n internal digital to analogue conversions for each analogue to digital conversion. The speed can be increased by incorporating an up-down counter in the controller and setting the direction of the count according to the relative values of V_D and V_A as measured by the comparitor. If the analogue input is increasing, the counter is incremented, and, if decreasing, it must be decremented from its previous conversion value. In general, the number of steps between conversions will be much less than with the cyclic counter, leading to potentially faster conversion times. The number of steps

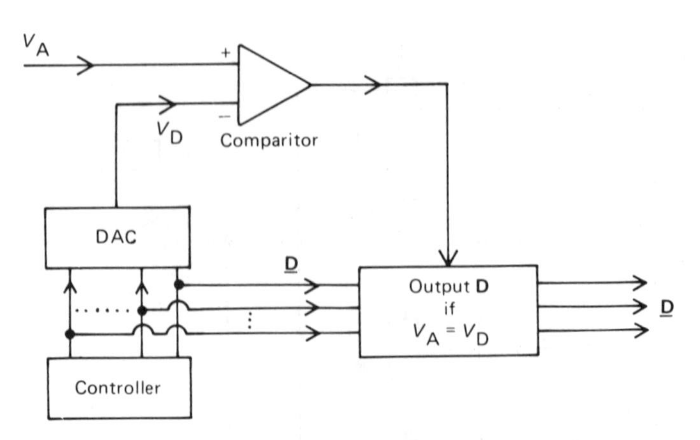

Fig. 7.5 A successive approximation analogue to digital converter.

between conversions will, however, vary with the rate of change of the analogue input. If the worst case conditions are allowed, where the analogue input swings from zero to its maximum value between consecutive conversions, then the up/down converter version will not be any faster than the cyclic controller.

An order of magnitude increase in speed can be obtained by replacing the counter with a processor which initially sets the most significant bit of the DAC to logical 1 and compares V_D wth V_A.

Initially all bits are at 0

If $V_D > V_A$ then the most significant bit of the DAC is reset to 0.
If $V_D < V_A$ then the most significant bit remains at 1.

The next bit is set to 1 and its final value determined by the V_D/V_A comparison. The procedure is repeated for each bit driving the DAC. The digital equivalent of the analogue input is the output of the controller after the least significant bit has been processed. This method requires n internal comparisons for each analogue to digital conversion, where n is the resolution in bits. The operating time in this method is independent of the value or rate of change of the analogue input.

Compare with 2^n operations for the counter-based converter.

Another commonly used technique is found in the dual slope integrating ADC. The essential components are shown in Fig. 7.6. The analogue input when applied

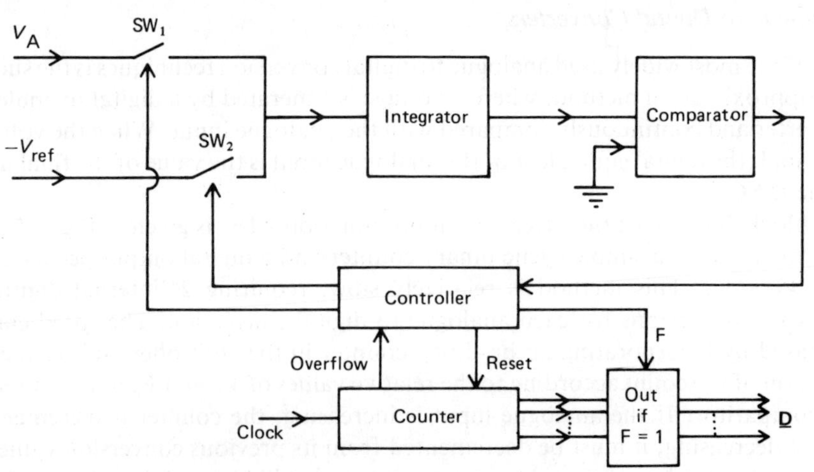

Fig. 7.6 A dual-slope integrating analogue to digital converter.

to an integrator starts an n-bit binary counter, driven from an accurate clock. The input voltage is continuously integrated until the n-bit counter overflows and returns to zero after 2^n clock pulses.

If the clock period is Δt and the analogue input is V_A, the output of the integrator when the counter returns to zero, will be

$$V_I = \int_0^{2^n \Delta t} V_A \, dt = V_A \, 2^n \Delta t \tag{7.2}$$

V_A is constant during conversion.

V_A is then switched out of the integrator, and a negative reference voltage ($-V_{ref}$) switched in, causing the integrator output to decrease. When the integrator output is zero (as detected by the comparitor) the counter is halted. If D is its final value, the integrator output can be expressed as

$$V_A \, 2^n \Delta t - V_{ref} D \Delta t = 0 \tag{7.3}$$

$$\text{Hence } V_A = \frac{V_{ref} D}{2^n} \tag{7.4}$$

So the final value of the counter is directly proportional to the digital equivalent of the analogue input V_A.

$\frac{V_{ref}}{2^n}$ is constant.

All the ADCs we have examined so far are relatively slow compared with DACs. The digital output is continuously built up and compared with the analogue input, and this is essentially a serial process. A parallel method is now feasible owing to the development of low cost microcircuits whereby the input voltage is compared with reference voltages for each and every possible digital output. These devices are called 'flash-converters' and their operation can be explained as follows.

Suppose we have a very simple converter that has 2-bits resolution and the analogue input voltage is in the range $0 \leqslant V_A \leqslant 4.0$V. The digital outputs 00, 01, 10

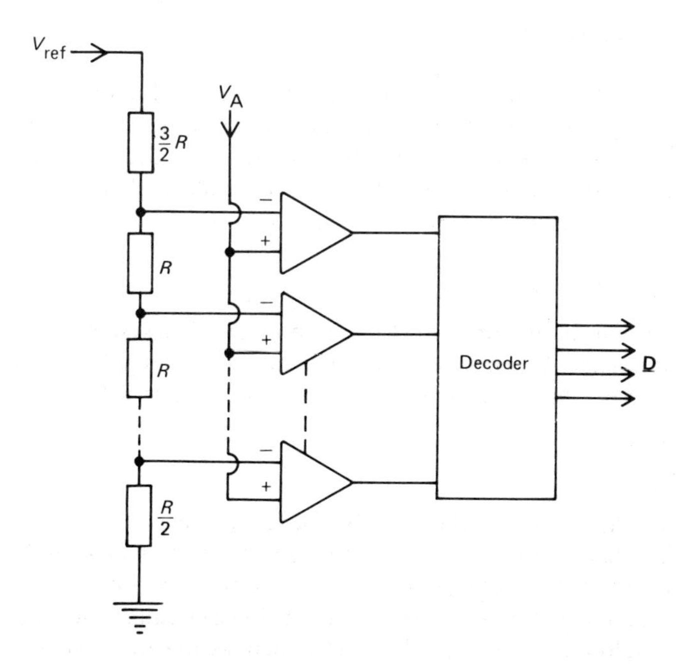

Fig. 7.7 A parallel or flash analogue to digital converter.

Table 7.2 A Flash Converter Output Decoder Table
Comparator at 0 implies $V_A < V_{ref}$ and at 1 implies $V_A > V_{ref}$

V_A	Comparitors 0.5V	1.5V	2.5V	Decoder output **D**
0V → 0.49V	0	0	0	0 0
0.5V → 1.49V	1	0	0	0 1
1.5V → 2.49V	1	1	0	1 0
2.5V → 3.99V	1	1	1	1 1

and 11 indicate whether V_A is closest to 0V, 1V, 2V or 3V, respectively. By choosing reference voltages at 0.5V, 1.5V and 2.5V and comparing the input V_A with each of these voltages in parallel, the digital equivalent of the input can be obtained. A comparitor output at 0 implies V_A is less than V_{ref}, and a 1 indicates V_A is greater than V_{ref}. The reference voltages can be generated from a precise resistor network as shown in Fig. 7.7 and the comparitor outputs must be decoded into pure binary as shown in Table 7.2.

Flash-converters are extremely fast as the comparisons are all performed at the same time. The reference input can be adjusted externally according to the maximum value of V_A to be encountered. An n-bit converter requires $2^n - 1$ comparitors so the hardware requirements for high resolution converters can become excessive. This method is therefore only used where high speed at low resolution is required. A typical application is the digitization of television pictures where the conversion frequency is of the order of 10MHz.

Typically 4-bits resolution.

Problems Arising from Logic Usage

The reader should now be familiar with the fundamental design principles for a digital system. The end-product of any successful design exercise is a working system, and this goal may be unattainable if certain practical precautions are not observed. In this final section, we will consider problems that arise when a logic system is actually built up on a printed circuit board.

Power Supply Decoupling

The wires and leads interconnecting logic components in a system will have a small but finite inductance. When a logic gate switches, a narrow spike of current flows in the power supply lines. A changing current in an inductor will generate a voltage

$$V = \frac{L\mathrm{d}i}{\mathrm{d}t} \qquad (7.5)$$

where L is the inductance of the wire. Suppose a current of 20mA flows for 100 nanoseconds when a gate switches. If the distributed inductance is 10^{-4} Henries, then a voltage of 20V is induced on the power rails. This voltage spike can cause serious malfunction and possible damage in the logic circuits. The induced voltage can be removed by connecting small radio frequency capacitors between the supply input and earth pins of each logic circuit. The capacitors decouple the power supply

$$V = L\frac{di}{dt} \simeq L\frac{\Delta i}{\Delta t} = \frac{10^{-4} \times 20 \times 10^{-3}}{10^{-7}} = 20\text{V}$$

and effectively short out the high frequency spikes. Typical values for the capacitors are 0.002μf to 0.1μf depending on the size of the integrated circuit package.

Pulse Reflections

A data highway or bus on a printed circuit when carrying information at very high frequencies can act as a transmission line. If a transmission line is not terminated with its characteristic impedance, reflections can occur from its ends. This may happen if the line is open circuit or terminated by a gate, and a pulse may be reflected back along the line, which will cause any logic connected to it, to switch a second time. This behaviour can be observed in high speed counters where reflections on the clock line can cause the counters to switch at twice the expected rate when fast narrow clock pulses are applied. If the pulse width is increased, there may be interaction with the reflections, and then erratic behaviour is observed.

Reflections can be eliminated by terminating the lines with their characteristic impedance by connecting a suitable resistance between the line and earth.

Typical values 50 – 300Ω.

Cross-Talk

Cross-talk can occur when signals running along adjacant tracks or wires are coupled together by mutual capacitance. This can lead to corruption of the data. Forward cross-talk occurs between adjacant tracks where the data signals run in the same direction. It is proportional to the signal edge speed (dV/dt) and the length of the coupled tracks. In general, problems usually occur when propagation distances exceed one metre.

Back cross-talk occurs when signals travel in opposite directions on adjacant tracks. It can cause serious problems over coupled lengths of a few centimetres.

Cross-talk can only be eliminated by modifying the layout of the circuit. It is reduced by increasing the spacing between tracks on a printed circuit board. It may be necessary to separate cross-talking tracks with a third track at ground potential, or in extreme cases, a ground plane may be needed. If cross-talk occurs between wires, a twisted pair of wires with one at ground potential will afford adequate isolation and in extreme cases, screened cable must be used.

For further details on printed circuit design refer to Scarlett, J.A. *Printed Circuit Boards for Microelectronics*, 2nd ed. *published by* (Van Nostrand Reinhold, 1984).

Ground is the same as earth potential.

Summary

Practical logic circuits can be divided into two broad groups. Bipolar logic uses the bipolar transistor, and MOS logic uses the unipolar transistor. Both types of logic can be fabricated as integrated circuits and within each group there is a range of products offering varying speed/power consumption trade-offs. Some bipolar and MOS circuits are incompatible, and additional interfacing circuitry may be required in mixed logic systems.

The number of connections to and from logic gates may have to be restricted so as not to overload the circuits. The manufacturers data must be adhered to when a logic system is built up.

A digital circuit will invariably have to be interfaced with an analogue system. Data converters may be used for this purpose. In general DACs are simpler and faster than ADCs, as analogue to digital conversion is a serial operation. Recent developments in integrated circuit technology have, however, made parallel ADCs a viable proposition for very fast but low resolution conversions.

Finally, the physical layout of a logic system has significant bearing on its reliability. The logic designer must decouple power supplies, terminate bus lines and ensure that the circuit layout is satisfactory, with adequate screening, to reduce cross-talk to a minimum.

Answers to Problems

1.1 (i) 10001101_2, (ii) 0.10111001_2, (iii) 10101.1101_2

1.2 (i) 57_{10}, (ii) 0.171875_{10}, (iii) 5.671875_{10}

1.3 (i) 10 bits, (ii) 4 octal places, (iii) 3 hex places, (iv) 13 bits

1.4 (i) 266_8, (ii) 1365_8, (iii) 57.6_8

1.5 (i) $B9C7_{16}$, (ii) $127C_{16}$, (iii) $FE.08_{16}$

1.6 (i) $100\,0111\,0010\,0011_{BCD}$, (ii) $110\,1001\,1001_{BCD}$, (iii) $100\,0011\,0101\,1001\,1000_{BCD}$

1.7 Correct data is

(i)	(ii)
0001	0110
1111	1000
1100	1100
0000	1110

1.8 A binary adder may be used, but each decimal stage must be checked to see if either a carry or an invalid BCD code is present (i.e., 1010 to 1111). If so, add 0110 to that stage.

Example	19 + 2 = 21
In BCD we expect	1 1001 + 10 = 10 0001.
In binary adder we get	1 1001 + 10 = 1 1011
But least significant group of 4	1011 is invalid
	∴ add 0110 to least significant BCD group.
	1 1011 + 0110 = 10 0001

1.9 Any code where adjacent numbers differ by only 1 bit can be used for position sensing.

1.10 In a weighted BCD code, each bit has a fixed numerical value. This is not the case in a non-weighted BCD code.

2.1 (i) NAND (ii) EX.OR (iii) NAND (iv) OR

2.3 (i)

ABC	F
000	1
001	1
010	1
011	1
100	1
101	1
110	1
111	0

(ii)

ABC	F
000	1
001	0
010	0
011	0
100	0
101	0
110	0
111	0

2.4

AB	F
00	0
01	1
10	1
11	0

$F = A \oplus B$

2.5

(i)

ABC	F
000	1
001	1
010	1
011	1
100	1
101	1
110	1
111	0

(ii)

ABC	F
000	0
001	0
010	1
011	1
100	1
101	1
110	0
111	1

(iii)

AB	F
00	1
01	1
10	1
11	0

2.8 (ii) $F = \overline{(A+B).\overline{\overline{A}.\overline{C}}}$

2.9 $F = \overline{A}\,\overline{B}C + \overline{A}B\overline{C} + A\overline{B}\,\overline{C} + ABC$

ABC	F
000	0
001	1
010	1
011	0
100	1
101	0
110	0
111	1

2.10

A B C D	Decision
0 0 0 0	0
0 0 0 1	0
0 0 1 0	0
0 0 1 1	0
0 1 0 0	0+
0 1 0 1	0+
0 1 1 0	0+
0 1 1 1	1*
1 0 0 0	0*
1 0 0 1	1+
1 0 1 0	1+
1 0 1 1	1+
1 1 0 0	1
1 1 0 1	1
1 1 1 0	1
1 1 1 1	1

* against A
+ against B

3.1 $F_1 = \overline{A}\,\overline{B}\,\overline{C} + \overline{A}\,\overline{B}C + \overline{A}B\overline{C} + A\overline{B}C + AB\overline{C}$

$F_2 = \overline{A}BC + A\overline{B}\,\overline{C} + A\overline{B}C + AB\overline{C} + ABC$

3.2 (i) $F_1 = \overline{A}\,\overline{C} + \overline{B}C$

$F_2 = \overline{A}B\overline{C} + A\overline{C}D + ABC + \overline{A}CD$

$F_3 = \overline{A}.\overline{B}$

(ii) $F_1 = (\overline{A}+C).(\overline{B}+\overline{C})$

$F_2 = (A+B+C).(\overline{A}+C+D)\,(A+\overline{C}+D)\,(\overline{A}+B+\overline{C})$

$F_3 = \overline{A}.\,\overline{B}$

3.3 $F_1 = \overline{\overline{\overline{A}B}.\ \overline{B\overline{C}}.\ \overline{\overline{A}CD}.\ \overline{A\overline{B}\,\overline{D}}}$

$F_2 = \overline{B.\ \overline{\overline{A}\,\overline{C}}}$

3.4 $F_1 = \overline{\overline{(\overline{A}+\overline{C})} + \overline{(B+\overline{C})}}$

$F_2 = \overline{\overline{(\overline{A}+\overline{B})} + \overline{(\overline{A}+C)} + \overline{(B+C)} + \overline{A+B+D}}$

3.5 1st canonical form: $F = \overline{A}B + A\overline{D}$

2nd canonical form: $F = (A+B).(\overline{A}+\overline{D})$

3.6 $W = \overline{\overline{A}.\ \overline{BD}.\ \overline{BC}}$

$X = \overline{\overline{\overline{B}D}.\ \overline{\overline{B}C}.\ \overline{B\overline{C}\,\overline{D}}}$

$Y = \overline{\overline{C\overline{D}}.\ \overline{\overline{C}D}}$

$Z = D$

3.7 $W = A$

$X = A\overline{B} + \overline{A}B$

$Y = B\overline{C} + \overline{B}C$

$Z = C + \overline{C}D$

3 EX.OR gates required.

3.8 Hazard-free circuit is: $F = AB + \overline{A}\,\overline{C}D + \overline{A}\,\overline{B}C + B\overline{C}D + \overline{A}\,\overline{B}D$

3.9 $F = BC + \overline{A}CD + \overline{A}C\overline{E} + \overline{B}\,\overline{C}\,\overline{D}E$

3.10 This problem does not have a unique solution. Answer depends on the cell specification.

F_1 \ AB	00	01	11	10
0	①	3	4	5
1	②	③	④	⑤

4.1 Stable states are ① , ② , ③ , ④ , ⑤

4.2

b_1	b_0	S	R	Q
0	0	0	0	0
0	1	1	0	1
1	0	0	1	0
1	1	0	0	0

4.3

$b_1 = J$	$b_0 = K$	Q
0	0	0
0	1	0
1	0	1
1	1	0

Timing problems eliminated by delaying counter clock.

4.4 9 gates.

4.5 If AB = 11 and internal state $F_1{}'\ F_2{}'$ is either 01 or 11, then $F_1\ F_2$ oscillates between 01 and 11.

4.6 Use a flow table to show that there is no transition from F = 1 to F = 0.

4.7

Q_2	Q_1	Q_0
0	0	0
0	0	1
0	1	0
1	0	1
1	1	0
0	0	1

4.8 $S_1 = A$, $R_1 = Q_1 Q_2 B + C$, $S_2 = A$, $R_2 = C$.
Protection circuit. AND each push button with the inverse of all the others before inputting to controller: e.g., $A_c = A\overline{B}\,\overline{C}$.

4.9

Q_2	Q_1	Q_0
0	0	0
0	0	1
0	1	0
1	1	1

4.10

Q_3	Q_2	Q_1	Q_0
0	0	0	0
0	0	0	1
0	0	1	1
0	1	1	1
1	1	1	1

5.1 $T_0 = I$; $T_1 = Q_0\ \overline{Q}_3$; $T_2 = Q_1$; and $T_3 = Q_2 + Q_0Q_3$. (No output logic required).

5.2 Output logic is $Z_3 = Q_2$, $Z_2 = Q_2\ \overline{Q}_0 + Q_2\ \overline{Q}_1 + \overline{Q}_2\ Q_1\ Q_0$, $Z_1 = Q_2 \oplus Q_1$ and $Z_0 = \overline{Q}_2\ Q_1 + \overline{Q}_2\ Q_0 + Q_2\ \overline{Q}_1\ Q_0$
Shift register version: use 4-bit shift register, inverted feedback and starting state 0000.

5.3 Driving logic for JKFFs $J_1 = Q_0$, $K_1 = \overline{Q}_0$, $J_0 = 1$, $K_0 = Q_1$
for TFFs $T_1 = Q_1 \oplus Q_0$, $T_0 = \overline{Q}_0 + Q_1$
for DFFs $D_1 = Q_0$, $D_0 = \overline{Q}_0 + \overline{Q}_1$

5.4 Either design a 4-state 3-bit counter to output R A G = 100,110,001,010,100, etc., or use a 2-bit pure binary counter and output logic: $Z_R = \overline{Q}_1$, $Z_A = Q_0$, $Z_G = Q_1\ \overline{Q}_0$.
Use the control waveform to drive the counter clock.

5.5 For faulty JKFF

J	K	$Q_t \rightarrow Q_{t+1}$
0	d	$0 \rightarrow 0$
1	d	$0 \rightarrow 1$
0	d	$1 \rightarrow 1$
1	d	$1 \rightarrow 0$

Ignore K and relabel J as T and sell as synchronous TFFs.

5.6 Driving logic: $J_1 = IQ_0$, $K_1 = 1$, $J_0 = I$, $K_0 = \overline{I} + Q_1$

System generates code sequence 00, 01, 11, 00 etc providing input I = 1. If I = 0 then reset to 00.

5.7

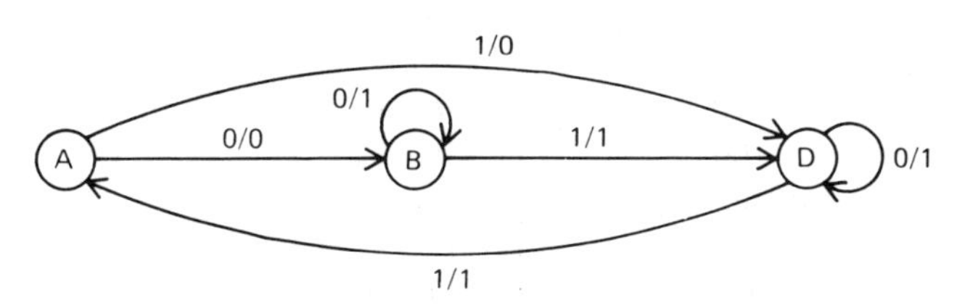

5.8 No unique solution. Obtain state transition diagram, state minimize to 4 internal states, obtain present/next state table and then design logic.

5.9 No unique solution. Proceed as in Problem 5.8. System minimizes to 4 internal states. Use an additional input push button (enter) to synchronize the input and clock the flip-flops.

5.10 No unique solution. This system minimizes to 7 internal states.

6.1 Multiplexer inputs

	D_0	D_1	D_2	D_3	D_4	D_5	D_6	D_7
(i)	0	0	1	0	1	1	0	1
(ii)	1	1	0	1	1	0	0	1
(iii)	0	0	0	0	0	1	0	0

6.2

(i)	0	1	0	$\overline{A}$	A	$\overline{A}$	1	0
(ii)	A	A	A	A	0	0	A	A

6.3

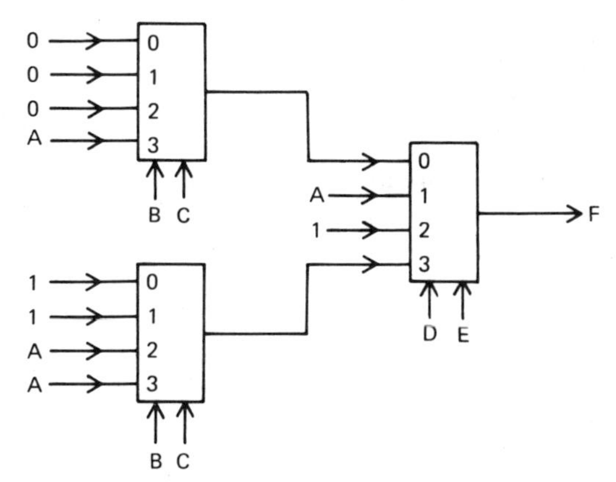

6.4 No unique solution. One simple method is to encode votes into 2 bits — for = 10, against = 01 and abstain = 00 and form an 8-bit word for the 4 voters. Address a 2^8 (256)-bit memory with the 8-bit word. Set each location in memory to 1 if its address word contains more odd bits (b_7, b_5, b_3, b_1) set to 1 than even (b_6, b_4, b_2, b_0).

6.5 Obtain present/next state table. Counter increments if I = 1 and halts if I = 0. Memory must have 6 address inputs (I and 5 output bits fed back) A 64 × 5-bit memory is required. Store present/next state table in memory. Feedback must be via DFFs in order to synchronize the system. Reset the system via clears on DFFs.

6.6 $4 \times 6 \times 3$ PLA.

6.7 No unique solution. F_1 and F_2 must be reduced to no more than 3 implicants between them. BC is common between F_1 and F_2.

6.8 No unique solution.

6.9

Fault	Test ABC	Result
a	010	$F = 0,\ F_a = 1$
b	100	$F = 0,\ F_b = 1$
c	undetectable	
d	100	$F = 0,\ F_d = 1$
e	000	$F = 0,\ F_e = 1$
f	010	$F = 0,\ F_f = 1$
g	or	$F = 0,\ F_g = 1$
h	100	$F = 0,\ F_h = 1$

6.10

Fault	Test ABC	Result
a	110	$F = 1,\ F_a = 0$
b	110	$F = 1,\ F_b = 0$
c	undetectable	
d	undetectable	
e	110	$F = 1,\ F_e = 0$
f	undetectable	
g	001 or 011	$F = 1,\ F_g = 0$
h	101 or 110, 111, 101, 100	$F = 1,\ F_h = 0$

Note. Faults a, b, and e are indistinguishable.

Index